GEOGRAPHY

A Study of Its Physical Elements

GEOGRAPHY

A Study of Its Physical Elements

Quentin H. Stanford and Warren Moran

TORONTO NEW YORK
Oxford University Press
1978

Canadian Cataloguing in Publication Data

Stanford, Quentin H., 1935-
   Geography: a study of its physical elements

Bibliography: p. 298
Includes index.
ISBN 0-19-540282-0

1. Physical geography.   I. Moran,
Warren, 1936-   II. Title.

GB55.S73   910'.02   C78-001409-X

Illustrations and maps by
Gus Fantuz, John Jakoi, and Margaret Kaufhold
Cover design and photographs by
Fortunato Aglialoro

ISBN 0-19-540282-0
12345-21098

Printed in Canada by
The Hunter Rose Company Limited

CONTENTS

1/ THE EARTH

Geographers are mainly interested in things on or close to the crust or outer shell of the earth. Yet some characteristics of the earth —the seasons, tides, day and night—can be understood only in relation to our solar system and the universe. It is necessary to see the earth as part of the universe before examining its surface.

The earth, the solar system, and indeed our galaxy—the Milky Way—are an extremely small part of the universe. Our largest telescopes have revealed an estimated 10 billion widely spaced galaxies, which are collections of stars and gas clouds. While no outer limits to the universe have been recognized, the total number of galaxies in the universe may be as many as 100 billion.

When we consider this enormous space called the universe, and the vast distances within it, the measurements used on earth no longer suffice. Instead, the common unit of measurement is the light year, that is, the distance light can travel in one year (the speed of light is approximately 299 460 km/s). Thus it has been estimated that the most distant observable galaxies are 10 billion light years away. Since the earth itself is estimated to be 4.5 billion years old, the light being observed by our telescopes from these distant galaxies started out long before our planet was formed.

A large galaxy such as our own Milky Way contains between 100 billion and 1000 billion stars. It has a diameter of 100 000 light years and varies in thickness between 5000 and 15 000 light years. The nearest galaxy is about 2 million light years distant. Like the Milky Way, it is disc shaped (similar to the galaxy shown in Fig. 1-1), with a bulge towards the centre.

Our sun is a star of medium size that occupies a position about halfway between the rim and the centre of the Milky Way. Like all members of our galaxy, the sun revolves around the galactic centre, completing one revolution every 200 million years at a velocity of 800 000 km/h.

The circumference of the sun is 100 times greater than that of the earth and its mass 3 000 000 times greater. By its gravitational force the sun rules the nine planets, their satellites, and also diverse meteors, comets, and asteroids that together make up our solar system. The sun is close to the centre of the orbit of the planets, which it also controls.

Like all stars the sun pours out energy, which we experience on the earth as light and heat. This energy is known as electromagnetic radiation. It is produced through nuclear reactions in the sun's interior (nuclear fusion) in which hydrogen is converted to helium under enormously high pressures and

Fig. 1-1. The spiral galaxy in the constellation of Ursa Major, seen through the 500 cm telescope on Mt. Palomar.

universe, although none has ever been observed. Judging from the enormity of this number there are likely to be many planets with conditions similar to the earth. Some form of organic life may exist on any of them. Thus, the evolution of life from primitive to more complex forms may have occurred often in our universe, but the great distances separating these planets make it unlikely that we shall ever prove their existence.

The earth, one of the smallest planets (Fig. 1-2), revolves around the sun at an average distance of 150 000 000 km. This revolution

temperatures. The amount of energy produced by this type of nuclear reaction is suggested by the following example. If hydrogen (which is found in water) could be converted to helium in nuclear reactors, only a few tonnes of water would be required to satisfy the world's energy requirements for a year.

Only a small portion of radiation from the sun's nuclear reactions reaches any planet. The amount that reaches a planet such as the earth depends on several things: its distance from the sun; its size; and the characteristics of its atmosphere. The amount of radiation reaching the earth is enough to sustain life as we know it. The conditions radiation creates on earth are thought to be unique in our solar system. Astronomers estimate that there may be as many as 10^{20} planetary systems in the

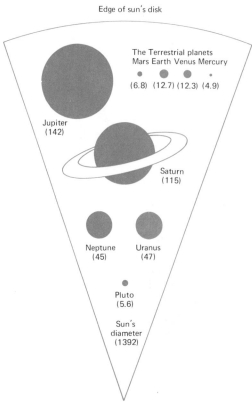

Fig. 1-2. Relative diameters of the Sun and planets. Figures give diameters in thousands of kilometres.

takes the form of a slightly elliptical orbit. It is balanced by two opposing forces: centripetal (gravitational) and centrifugal. On the one hand, the sun has a gravitational pull that keeps the earth and its sister planets from whirling off into space. Counteracting this pull—so that the sun does not draw them into itself—the orbiting planets develop a centrifugal effect that causes them to fly away from the centre of revolution. The balance between these two forces determines the velocity and orbital path around the sun of the earth and other planets.

The earth has an orbiting satellite of its own, the moon. It revolves around the earth every 27 d and 8 h at a distance of about 386 400 km. Only one side of the moon can be seen from the earth. (Spacecraft have now made available photographs of the other side.) This is the result of two related factors: first, the moon takes as long to rotate once on its axis as it does to complete one revolution around the earth, and second, both the revolution and rotation of the moon occur in the same direction. The investigation of the moon's surface characteristics and internal structure is beginning to help us better understand some of the occurrences that formed our own planet. Unlike the earth, the moon has no atmosphere. Therefore its surface features have not been subjected to erosion. Marks left by events of the past billions of years, such as bombardment by meteors or volcanic eruptions (Fig. 1-3), are still visible on the moon's surface. Similar events probably occurred on the earth, but their effects have been worn away.

The earth has a diameter of approximately 12 800 km. Its shape is usually described as spherical, but not a perfect sphere. As a result of the earth's rotation, the region around the equator bulges out slightly, while the regions around its poles are slightly flattened. (The equatorial diameter is approximately 12 763 km, which is 44 km greater than the polar diameter.) The shape of the

Fig. 1-3. An oblique view of Schroeter's valley and, to the left of the valley, the crater Aristarchus (photographed by the Apollo 15 Command module in lunar orbit). The crater is approximately 35 km in diameter.

earth is more correctly described as an irregular *oblate spheroid*. Other minor irregularities on the outer part of the earth's surface include bulges and depressions in the crust. The variations range from about 8 km above sea level atop the highest mountains to about 11 km below sea level in the deepest parts of the ocean. Though they appear large when viewed by people, in terms of the earth's total size they detract little from the general smoothness and roundness of the earth.

Spread over this irregular surface are an estimated 1 323 000 000 km³ of water covering some 71 per cent of the earth. The average depth of this water is approximately 3750 m, while the average height of the higher parts of the crust or the land masses remaining above water is approximately 750 m. If the surface of the earth were smooth, with all irregularities removed, the entire globe would be covered by about 2400 m of water.

Fig. 1-4. The principal planets.

	Name	Mean Distance from Sun 10⁶km	Inclination of Axis	Period of Revolution	Orbital Velocity km/s	Period of Rotation	Diameter 10³km	Mass, Relative to Earth	Number of Moons
The Terrestrial Planets	Inner planets			Days					
	Mercury	58	28°	88	47.9	58 d 17 h	4.9	0.06	0
	Venus	108	3°	225	35.0	243 d	12.2	0.81	0
	Earth	150	23°27′	365.25	29.8	23 h 56 min	12.7	1.00	1
	Mars	228	24°	687	24.1	24 h 37 min	6.7	0.11	2
The Great Planets	Outer planets			Years					
	Jupiter	779	3°5′	12	13.1	9 h 50 min	142	318	13
	Saturn	1430	26°44′	29.5	9.6	10 h 14 min	115	95	10
	Uranus	2870	82°5′	84	6.8	10 h 42 min	47.4	15	5
	Neptune	4500	28°48′	165	5.4	15 h 48 min	44.6	17	2
	Pluto	5900	?	248	4.7	6 d 9 h	5.6(?)	0.9	0

STUDY 1-1

1 What is the equatorial circumference of the earth?

2 Find the height of the highest mountain on earth and the depth of the deepest part of the ocean. If the earth was reduced to a large globe 60 cm in diameter—with all the irregularities of its surface shown to scale—how great would be the difference between these two extreme points? Would it be possible for the naked eye to detect such irregularities on a globe of this size?

3 Everest is the highest mountain relative to sea level. Would the summit of this mountain represent the farthest point from the earth's centre? Explain.

THE EARTH'S GRID

Before we begin to examine the earth and its surface, it is necessary to understand the problem of precisely locating places on the surface of the earth. To determine the precise location of any place, it is necessary to use a grid or network of imaginary lines intersecting at right angles. The need for such a grid is clear if one tries to describe the location of a point on the surface of an unmarked sphere such as a ball. It is impossible to do this without drawing lines, which of course must have a beginning, or point of reference. Since the earth is a rotating sphere, its most convenient points of reference are the two ends of the axis—the poles. (The *axis* is an imaginary line from pole to pole around which the earth rotates.)

Using the poles as reference points, two base lines were created. The first, the *equator*, is a line joining all points on the earth's surface that are midway between the poles; it divides the earth into equal halves. The second could have been any line extending from pole to pole. However, one was chosen by international agreement in 1884 to run through the Royal Observatory at Greenwich, England. It is known as the *prime meridian*.

Using these lines as a basis, a grid of imaginary lines was developed. The lines running east-west are called *parallels of latitude*. They indicate locations north or south of the equator, and are parallel to the equator. The lines running north-south are called meridians of longitude. They indicate locations east or west of the prime meridian, and extend from pole to pole like the prime meridian. They are called *meridians of longitude*. These two sets of lines intersect at right angles.

In order to identify lines in the grid, each line is given a numerical value. The most suitable means of identification on the surface of a sphere is a system of measurement based on angles. One circle or complete rotation = 360° (degrees); 1° = 60′ (minutes) and 1′ = 60″ (seconds). This concept is easier to understand when it is applied to a diagram such as Fig. 1-5. Thus the lines parallel to the equator have the value of the angle that is formed when a line is extended from any point on the parallel to the centre of the earth and back along the equatorial plane (LAXB in Fig. 1-5 is 30° and identifies the parallel 30°N). Parallels of latitude are identified by their angular distance north or south of the equator. Similarly, a line extending from pole to pole has the value of the angle formed when a line is drawn from any point on the meridian to the axis and back to a point on the prime meridian at the same latitude (LEXB or LCYD in Fig. 1-5 are both 45° and identify the meridian 45°W). Thus meridians of longitude are identified by their angular distance east or west of the prime meridian.

The location of any point on the earth can be described by referring to the parallel of latitude and the meridian of longitude passing through that point. If this is expressed in degrees, minutes, and seconds, the location will be accurate to within a few metres. To avoid confusion, the hemisphere in which the measurement is being made must be indicated by noting after the value of the latitude whether the point is *north* or *south* of the equator, and after the value of the longitude

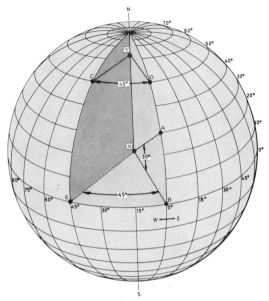

Fig. 1-5. A portion of the earth has been removed to show the angles used to identify the parallels and meridians.

whether it is *east* or *west* of the prime meridian. Because minutes and seconds are rather awkward to use, latitude and longitude can be expressed as decimal parts of the degree. For example, latitude 40°30′N becomes 40.5°N.

STUDY 1-2

1 The earth's polar circumference (360°) is 40 025 km. Any part of this circumference is referred to as an arc (a curved line) because the earth is a sphere. What is the length of the arc made on the surface of the earth by a degree of latitude? Will that distance vary on different parts of the earth's surface? Can the same calculation be made for a degree of longitude? Where is the length of a degree of longitude the same length as a degree of latitude? In each case explain your answer.

2 Prepare a sketch similar to Fig. 1-5 showing the angles that identify the position 20°N and 80°W.

3 Using an atlas, answer the following:

a Find to the nearest degree the latitude and longitude of Cairo, Melbourne, Cape Horn, Pitcairn Island, and Cape Race.

b What is located in each of the following positions: 24°N 32°45'E; 35°50'N 140°E; 27°S 109°W; 75°N 100°W; 43°40'N 79°25'W; 24°30'N 54°20'E?

c Using latitude only, find the approximate distance in kilometres from Cape Horn to the northernmost point on Baffin Island, from Lagos to Algiers, and from Tokyo to Melbourne.

4 In Fig. 1-6:

point A is Greenwich 52°N 0° longitude;
point X is the centre of the earth;
point B is on the opposite side of the earth from point P, D from E, and M from A.

CXE = 35°
CXH = 65°
CXF = 100°
CXP = 27°

Find the latitude and longitude of points B to N.

5 From 13°N 150°W move latitudinally 25° south and longitudinally 40° west. State the latitude and longitude of your new position.

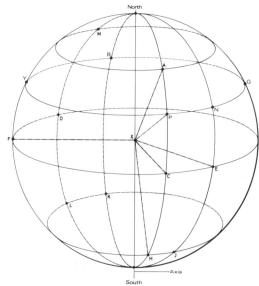

Fig. 1-6.

THE MOTIONS OF THE EARTH

It is difficult to comprehend that our planet is not hanging immobile in space or, at best, moving very slowly. We live with an illusion of terrestrial stability. It is much easier to believe that the stars, other planets, and moons, are moving about us while we are stationary. Until the discoveries of Copernicus and Galileo in the sixteenth and seventeenth centuries, people held to the theories expressed in the *Amalgest*, an encyclopedia of astronomy written by the Greek, Ptolemy, around 150 A.D. This work stated that the earth was stationary and the sun and other planets circled around it. We now know that this geocentric view is wrong and the earth's movements are far more complex. Two of the earth's movements that have an influence on its surface are the rotation of the earth on its axis and the revolution of the earth around the sun.

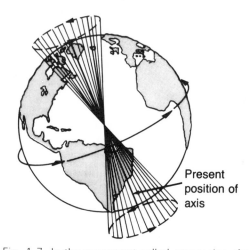

Fig. 1-7. In the movement called *precession*, the tilted axis of the earth swings in a tight circle to trace a double conical figure. The effect is very similar to the motion of the axis of a spinning top. (The effect of nutation cannot be seen in this diagram.)

Rotation

The earth rotates on its axis from west to east, completing one rotation approximately every 24 h 56 min. The axis itself does not have a fixed orientation in space. Rather it swings slowly around, completing a full circle once every 26 000 years (Fig. 1-7). During this movement, known as *precession*, the axis does not describe a perfect circle because the different gravitational pulls of the sun and moon cause a wobbling effect called *nutation*.

Because of the earth's rotation, almost every point on its surface has times of darkness and sunlight—night and day. One effect of rotation is obvious to us in the apparent movement of the sun and other stars. We think of the sun as 'rising' in the east and 'setting' in the west, but it is actually the rotation of the earth towards the east that makes the sun seem to move. In addition to causing day and night, the earth's rotation also has an important influence on the general circulation of air and water on the earth's surface.

Revolution

While rotating on its axis, the earth also travels around the sun in an elliptical orbit. This movement is known as *revolution*. The earth completes one revolution every 365.242 rotations, that is, in 365.242 d. Because its orbit is an ellipse and the sun is located at one focus, the earth's distance from the sun varies from a minimum of 147 million kilometres on 3 January (called *perihelion*) to a maximum of 152 million kilometres on 4 July (called *aphelion*). During its orbit the earth remains on an imaginary plane, as if it were travelling on a flat surface. Its axis forms an angle of 66°33′ to this *plane of orbit* (more correctly known as the *plane of the ecliptic*).

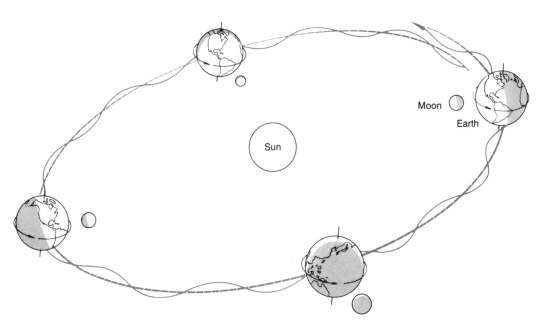

Fig. 1-8. The path of the earth-moon system as it revolves around the sun. The break in orbit shown on the diagram indicates that this system does not quite retrace its orbit in the following year.

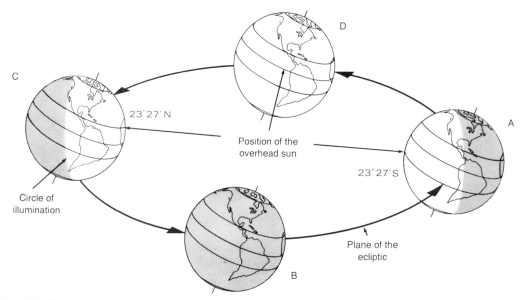

Fig. 1-9. The seasons occur because the tilted axis of the earth maintains a constant orientation in space as the earth revolves around the sun.

This means that the axis is 23°27′ off the true perpendicular. The position of the axis stays at this angle and always remains parallel to its previous positions. This is known as the *parallelism of the axis* (Figs. 1-8 and 1-9).

The earth is involved in further motions. As mentioned earlier, the sun revolves around the centre of the galaxy, and the galaxy itself is moving through space. All of these motions occur at very great speeds. The speed of the earth's rotation at the equator is 1675 km/h. The earth revolves around the sun at 106 300 km/h, and the sun moves around the centre of the galaxy at 800 000 km/h.

The most important effects of the earth's revolution are seasonal differences in temperature and variations in the length of day and night at different points on the earth's surface (discussed in Chapter 3). Of course, it is important to emphasize that if the axis was not inclined to the plane of orbit at an angle of 66°33′, the revolution of the earth would not produce the above variations.

STUDY 1-3

1 Draw the earth with a diameter of 5 cm and show the poles, the axis, and the equator. Draw the plane of orbit, showing the 66°33′ angle (approx.) formed between the plane and the axis. The plane of orbit will pass through the centre of the earth, and the 66°33′ angle can be shown with either the northern or the southern half of the axis. On one side of the earth show the parallel rays of the sun (parallel to the plane of orbit) striking the earth. Shade in that part of the earth in darkness. How much of the earth is in darkness? What parts of the earth's surface will spend a complete rotation in darkness or in light? Using an arrow, show the direction of the earth's rotation. How do we know that this is the direction of rotation? How many days of the year does this diagram represent?

2 In 1582 Pope Gregory XIII created our present calendar. In the Gregorian calendar, every year that is evenly divisible by four is a

leap year, except century years that are only leap years if they are divisible by 400. The accumulating error in this calendar is about one day every 3300 years. In what way is the Gregorian calendar related to the fact that the earth requires 365.242 d to complete one revolution of the sun?

3 If the speed of rotation of the earth is 1675 km/h at the equator, what will be the speed of rotation at the poles (90°) and at 40° latitude? (The circumference of the 40th parallel is approximately 30 838 km.)

DAY AND NIGHT

Although the sun is a sphere, its rays are virtually parallel when they reach the earth because of the size of the sun and its distance from the earth. As the earth rotates on its axis, one half is lit by these parallel rays while the other half is in darkness. Dividing the two halves is a line known as the *circle of illumination* (also called the *terminator*). Almost every place on the earth experiences a period of daylight and one of darkness as the earth completes one rotation. If the earth's axis formed an angle of 90° to the plane of orbit, the circle of illumination would pass through both poles. Every place on earth would then experience almost equal periods of daylight and darkness. However the axis is tilted at a constant angle of 66°33′ from the plane of orbit. This results in varying lengths of daylight, not only from place to place on the earth, but also at the same location over the course of a year.

The change from light to dark at the circle of illumination is not an abrupt one. A diffused light, known as *twilight*, occurs after sunset and before sunrise. It is caused by the sun's rays being scattered and reflected by dust and moisture particles in the atmosphere. At sunset this diffused light gradually weakens and then disappears when the sun is more than 18° below the horizon. The reverse applies before sunrise.

The Equinoxes, Spring and Fall

Equinox means the time of equal night, from the Latin *aequinoctium*. In Fig. 1-9, B and D represent the earth on 23 September and 21 March. On these two days the circle of illumination passes through both poles and days and nights are equal (12 h each) over the entire earth.

The Solstices, Winter and Summer

Solstice means the time when the sun seems to stand still, from the Latin *solstitium*. In Fig. 1-9, A and C represent the earth on 22 December and 22 June. Examination of A will show that the earth's plane of orbit and its position in orbit are such that the sun's rays strike directly 23°27′ south of the equator and therefore the circle of illumination does not pass through the poles. Instead, it extends from the western edge of the Arctic Circle on the side facing the sun to the eastern edge of the Antarctic Circle on the side away from the sun. As a result, places in the northern hemisphere experience a much shorter period of daylight—or no daylight at all for places north of the Arctic Circle. The opposite occurs in the southern hemisphere. In C, however, although the angle between the axis and the plane is unchanged, the sun's rays are striking most directly north of the equator at 23°27′N. On this day the North Pole is constantly in sunlight. A person standing at the pole would observe the sun moving counter-clockwise in a circle at a constant elevation of 23°27′ above the horizon. In the northern hemisphere, 22 December is called the winter solstice, and 22 June is called the summer solstice. The reverse is true for the southern hemisphere.

STUDY 1-4

1 Using Fig. 1-9 as a guide, make a similar sketch but change the point of view to show

the earth and the circle of illumination as they would appear from directly above the North Pole. (The circle of illumination should be shown as a curved line.)

2 Make a table to show the variation in the length of night and day on the following dates —23 September, 22 December, 21 March, 22 June—and at the following latitudes— 80°N, 40°N, 0°, 40°S, 80°S. Some of the figures required in this table will have to be estimated. Why is it not necessary to use more than these four days to illustrate the yearly variation in day and night at different latitudes?

3 Develop a diagram to show that the length of twilight is dependent on latitude. Explain how the appearance of the terminator on earth would differ from its appearance on the moon.

4 Do all planets have an equator? Do all planets have 'arctic circles' and 'tropics' (see Fig. 1-4)? Are the latitudes of these lines (if they have them) similar for all planets? Why?

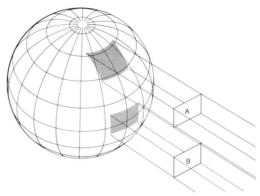

Fig. 1-10.

THE SEASONS

The pattern of day and night, of light and darkness, is experienced by everyone and often determines our activities. Yet there is another related phenomenon that has much wider implications—the variations in temperature that are experienced everywhere on the earth's surface in each 24 h cycle and in most places from one season of the year to another. The relationship between length of day (or duration of sunlight) and seasons is relatively straightforward: the longer the period of sunlight, the greater the amount of solar radiation a place will receive.

A second factor in the development of seasons is the variation in the intensity of the sun's rays at different times of the year. Variations in intensity result from variations in the angle at which the sun's rays reach the surface of the earth. In Fig. 1-10, A and B represent two equal amounts of solar radiation

striking the earth. The rays (A) in the higher latitudes strike the earth at a more acute angle and deliver less energy or heat because they must pass through a thicker layer of absorbing and reflecting atmosphere and heat a larger area of the earth's surface than the most direct rays (B). As the angle made by the sun's rays becomes more acute in the higher latitudes, the amount of heat received at the earth's surface decreases.

The inclination of the earth's axis causes the variation in the angle at which the sun's rays strike the earth as it rotates and revolves in its orbit. If the earth's axis were not inclined, the angle made by the sun's rays at any point on the earth's curved surface would remain the same all year round, and there would be no significant seasonal variation in temperature. In fact all days everywhere on the earth would be 'equinox' days. To polar observers the sun would occupy a perpetual sunset position and people on the equator would receive the sun's vertical rays every day at noon.

During the solstices the mid-day sun is overhead at 23°27' either north or south of the equator because of the earth's 66°33' inclination (this is shown in Fig. 1-9). The 23°27' parallel in the northern hemisphere is known as the *Tropic of Cancer* and the 23°27' parallel in the southern hemisphere is called the *Tropic of Capricorn*. When the noon sun is overhead at the Tropic of Cancer, the angle between the sun's rays and the horizon is much greater (more direct) at any given latitude in the northern hemisphere than at a corresponding latitude in the southern hemisphere. The greater amount of energy received in the northern hemisphere at this time establishes the warm—or summer—season, while it is winter in the southern half of the world. The equinoxes are intermediate times when the amount of energy received in both hemispheres is approximately the same because the sun at noon is directly overhead at the equator. Figure 1-12 shows the altitude of the sun, that is, the angle between the sun's rays and the horizon, at noon at various latitudes in the northern hemisphere. Figure 1-11 illustrates the angle on 22 December at various latitudes.

STUDY 1-5

1 Make diagrams similar to Fig. 1-11 showing the angle formed between the sun's rays and the horizon at the equinoxes and on 22 June.

2 Explain the difference in seasons on 22 June between northern China and Tasmania. Both of these places are in the 40-50° latitudinal range. In your answer estimate the actual difference in the angle of the noon sun and the duration of daylight in the two areas.

3 Hardly any seasonal differences in temperature occur at the equator and for a considerable distance on either side of it. At the poles, however, seasonal variations are large. Explain.

4 Calculate how many degrees warmer Miami is than Arctic Bay in both summer and winter, and explain why the difference is so much greater in one season than the other.

Fig. 1-11. Winter solstice.

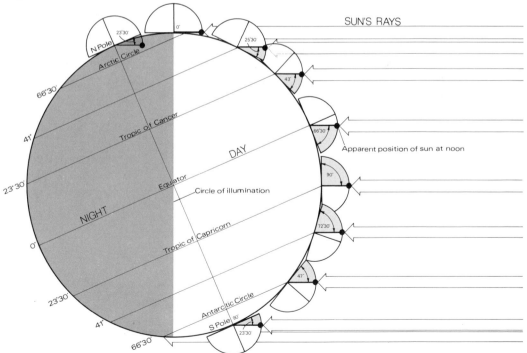

Fig. 1-12.

Latitude	22 December	21 March & 23 September	22 June
90°	below horizon	0°	23°27′
66°33′	0°	23°27′	47°
41°	25°30′	49°	72°30′
23°27′	43°	66°33′	90°
0°	66°33′	90°	66°33′

Why is the annual range of temperature greater at Arctic Bay than it is in Miami?

	January	July
Arctic Bay (73°N 84°W)	-29°C	6°C
Miami	20°C	28°C

TIME

The measurement of time is based on the apparent movement of the sun, that is, the sun appears to make one complete circuit of the earth every 24 h. When the sun appears to reach its highest point in the sky every 24 h, it is directly above the meridian of the observer; this time is known as *apparent solar noon*. A time system based on the actual observation of the sun, for example time shown by a sundial, is called *apparent solar time*.

Unfortunately the sun is a bad timekeeper. Sometimes it runs slow, sometimes fast, with a total range over a year of approximately 30 min from one extreme to the other. During the course of a year, the amount of time the sun runs fast is equal to the amount of time it runs slow, and therefore each rotation during the year averages out to exactly 24 h. The system of time based on such an average is known as *mean solar time*. The difference between apparent and mean solar time, as shown on the analemma in Fig. 1-13, is referred to as the *equation of time*. For example, on 15 October the analemma shows the sun is running 14 min fast, which means that apparent solar noon on this day will occur 14 min before mean solar noon.

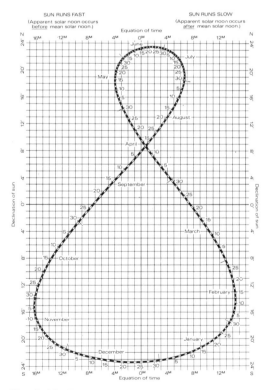

Fig. 1-13. The analemma allows both the sun's declination and the equation of time to be estimated for any date in the year.

Until the 1880s most localities used mean solar time, which was known as *local time*. According to this system any place on the same meridian would have the same time, but east or west of that meridian small adjustments were made for every change of longitude. Since the earth turns through 360° every 24 h, when it is noon at one meridian it is one hour later at the meridian 15° to the east and one hour earlier at the meridian 15° to the west. Obviously in the days when most cities and towns kept their own time, small differences in time occurred between places located on different meridians of longitude. For example, if it was noon at Washington, D.C., the time in New York City would be 12 12 and in Richmond, Virginia, 11 58.

Outside North America there were many variations in the strict application of local time. In many European countries the time for the whole country was determined by the local time of the capital, and time changes occurred only when international boundaries were crossed.

Standard Time

The system of local time created many problems, particularly after the development of the telegraph in 1845 and the more rapid forms of transportation in the late nineteenth century. For example, most American railroads adopted their own system of standard time to overcome the many time changes that occurred along their routes. Railroad time remained the same for long distances on each route, but it usually differed from local time in every town through which the railroad passed. Where two or more railroads went through the same town, it was necessary to distinguish between several different times— one for each of the railroads as well as the town's own local time. It has been estimated that in the 1880s in the United States there were as many as five different time systems in some localities, with the railroad companies

following 53 different systems across the entire country.

The world's commercial countries needed a solution to this confusing state of affairs. In 1884 an international congress was held in Washington, D.C. It recommended the adoption of a world-wide system of time zones based on standard time meridians spaced at intervals of 15°, as shown on Fig. 1-14. Subsequently most countries adopted Standard Time based on the meridian of longitude of Greenwich, England—the prime meridian. This meridian was chosen for practical reasons. It had been used by sailors as a basis for navigation since the first nautical almanac was produced in 1767. Most of the world's shipping companies had the Greenwich meridian on their charts, and North American railroads were already using a time-zone system based on Greenwich. At an international conference in 1883 the meridian passing through Greenwich Observatory was named Prime Meridian and numbered 0° longitude. Some of the important characteristics of Standard Time can be deduced with the help of Fig. 1-14 and the questions in Study 1-6.

STUDY 1-6

1 How many time zones is the world divided into (Fig. 1-14)? Explain (a) why this number was chosen and (b) the time change that takes place between adjacent zones. Which country has the greatest number of time zones?

2 The time in each zone is based on the local time of the meridian, known as the *standard meridian*, that bisects the zone. The standard meridian of the Greenwich time zone is 0°. What are the standard meridians for the principal time zones of North America?

3 An examination of a map of time zones will show that the boundaries of most time zones do not follow the meridians but take an irregular course in the vicinity of the correct meridian. What are some of the reasons for these

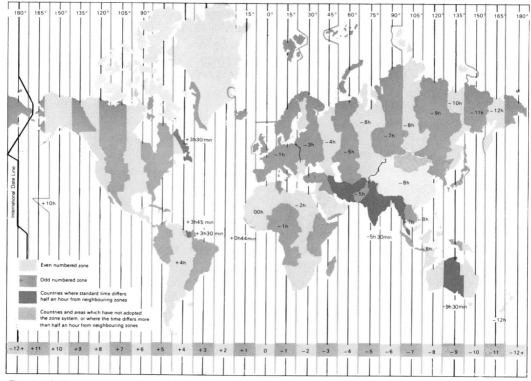

Fig. 1-14. Time zones of the world.

deviations? Give some examples to illustrate your answer.

4 Cape Town in the Republic of South Africa and Helsinki, Finland, are thousands of kilometres apart, yet their time is the same. Regina and Yorkton in Saskatchewan are less than 250 km apart, but there is a time difference of one hour between the two places. Explain these two situations.

5 From Fig. 1-14 it can be seen that some areas have times that are faster or slower than Greenwich by multiples of a half-hour. Can you explain why this is so?

6 A traveller leaving Wellington, New Zealand, at 15 00 on 8 January took 22 h to reach Rome, Italy. What was the time in Rome when he arrived?

7 What is daylight saving time? Why is it used? Do all parts of a time zone have the same need for daylight saving time?

8 Why have the short forms a.m. and p.m. been used in connection with time?

The International Date Line

A practical problem involving time was first recorded in the early sixteenth century when Ferdinand Magellan sailed his flagship *Trinidad*, together with the *Vittoria* and the *Concepion*, westward around the world, passing the tip of South America on his way. When the crew arrived back in Spain three years later, they found their calendar a day behind the local one. Setting their time by the sun as they travelled west, these explorers had lost 24 h after circumnavigating the globe.

Another illustration of the same problem

can be seen by examining the map of time zones (Fig. 1-14) and noting the difference in hours from the Greenwich time zone to the time zone in which 180° is the standard meridian. Counting east from Greenwich, the time in this zone is 12 h ahead of Greenwich, while counting west it is 12 h behind. Thus, in the 180° time zone, that part to the west of the 180° meridian (the Asian side) is 12 h fast, while that part to the east (the American side) is 12 h slow, and the difference in time between the two is 24 h, or a full day. If it is 15 00 on Sunday on the American side, it is 15 00 on Monday on the Asian side of this meridian.

Because of this unusual situation, and to avoid confusing travellers, the 180° meridian was designated in 1884 as the International Date Line. When this meridian is crossed, travellers must change their calendars, repeating a day when crossing it from west to east and losing a day going in the opposite direction.

1 What was the significance of the arrival date of Magellan's ships in Spain in terms of the direction in which they circumnavigated the globe? Was the length of time of the voyage important in determining the difference of time between the ship's calendar and the local one? Explain.

2 The International Date Line was established in 1884 after the Greenwich meridian had come into widespread use. What was so fortunate about the location of the 180° meridian? Why does the Date Line not exactly follow the line of this meridian?

3 If it is 15 00 on 28 July at Greenwich, what is the time and date at 175°W, 175°E, 120°E, 150°E, and 135°W?

4 If it is 22 00 on Friday, 13 June in Hawaii, what is the date and time in Sydney, Australia?

5 Is there any point in time when the same calendar day exists everywhere in the world?

6 At any specific place on earth we know that one day lasts 24 h. How long will any one specific calendar day continue to exist somewhere in the world regardless of location? What is the significance of the large bends in the date line to the duration of a calendar day?

THE BIOSPHERE

To this point we have been concerned with the earth and its planetary characteristics. For the most part we said nothing about life on earth. It is not difficult to become immersed in the study of landforms, climates, or vegetation for their own sake, and neglect their interrelationships with people. One important purpose of physical geography is to examine the relationship between people and their natural environment. This relationship involves our use of and effect on those elements of nature crucial to our existence and material culture. Some of these elements are the air we breathe, the soil we plant, or the rocks we mine.

The term biosphere has been used for some time to refer to that part of the earth in which life exists. The biosphere is the life layer, a shallow but extremely complex zone (Fig. 1-15) that includes the lower levels of the air (atmosphere) and the upper levels and surfaces of the land (lithosphere) and oceans (hydrosphere). Thus the biosphere occurs at the contact between these spheres.

What makes the biosphere so special? First, it receives a substantial supply of energy from the sun; second, it contains large quantities of liquid water; and third, there are contacts between the three states of matter—solid, liquid, and gas. These characteristics are the conditions necessary for life and they do not occur in any other part of our solar system.

The living organisms (plant and animal) that inhabit the biosphere interact with each

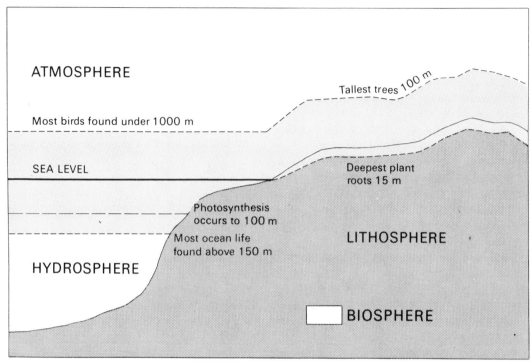

ATMOSPHERE

Tallest trees 100 m

Most birds found under 1000 m

SEA LEVEL

Deepest plant
roots 15 m

Photosynthesis
occurs to 100 m

Most ocean life
found above 150 m

LITHOSPHERE

HYDROSPHERE

BIOSPHERE

Fig. 1-15. The biosphere is a thin layer encircling the earth. It sustains a hierarchy of millions of planets and animal species, all interconnected and dependent on one another for life support.

other and their environment. The sum of all these interacting components is called an *ecological system* or, more commonly, an *ecosystem*. An ecosystem includes not only living matter but all the physical features that make up the environment. It is important to note that ecosystems can exist at a great variety of different scales in the biosphere. A drop of pond water containing such minute life forms as protozoa is an ecosystem; so is the pond itself with its complex range of life forms. At the highest level, the entire earth's biosphere is an ecosystem. Thus small ecosystems exist within larger ones, and one ecosystem is related to others in a variety of complex ways. Relationships between living and non-living parts of an ecosystem are not simply one-way. The physical environment influences life but life in turn affects the physi-

cal conditions of the ecosystem. This interaction is one with considerable implications for people. Our capabilities for modifying and changing nature are awesome and our impact on the biosphere in just the past few decades alone has been dramatic. The consequence of human actions on the various ecosystems may not always be immediately noticeable but it is extremely important for us to undertand that changes are occurring. Physical geography contributes to this by examining the physical components of the biosphere which, in essence, is the object of this book.

Energy and Ecosystems

Solar energy is crucial to the existence of life

in all ecosystems. Through the process of photosynthesis solar energy is used by all green plants to produce organic molecules that are necessary for the growth of the plant. Plants are the *producers* of the ecosystem. But not all the energy gained by photosynthesis appears as plant tissue, for much is lost through respiration and the life processes that keep the organism operating. After the losses through respiration are subtracted from the energy reaching the earth's surface, it is estimated that less than 1 per cent actually ends up as plant material (known as *net primary production*). This small per cent represents approximately 150 to 200 billion tonnes of organic material per year for the earth as a whole, including the food required by all life (plant and animal) in the biosphere. Net primary production varies greatly from one type of ecosystem to another. Those with the highest annual production include tropical forests and freshwater swamps (approx. 2 × 10^6 kg/m³). Agricultural land and middle latitude grasslands have a lower annual productivity (approx. 6 × 10^5 kg/m³), while desert and polar ecosystems are very low 10^4 kg/m³).

Plants, therefore, make their own food. All remaining organisms in the ecosystem (referred to as *consumers*) are ultimately dependent on plants and the process of photosynthesis for their existence (see Fig. 1-16). Some animals (herbivores) are primary consumers and live by feeding directly on plants. Secondary consumers (carnivores) feed on the primary consumers, and in most ecosystems still higher levels of consumers are evident. Such a series of levels (called *trophic levels* and also referred to as a *food chain*) may be thought of as an energy flow system in which energy is lost to the system at each level in the chain. As shown in Fig. 1-17 the losses at each level in the food chain are considerable, a factor with great implications for human food production. For example, where animals are raised for meat, the amount of

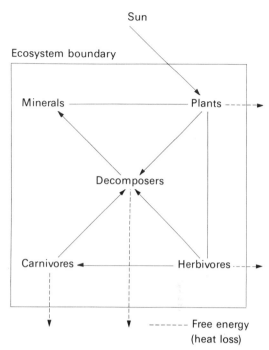

Fig. 1-16. Within an ecosystem, energy is used to produce material by photosynthesis. The existence of almost all other forms of life—consumers and decomposers—depends on plants and the process of photosynthesis.

food produced in a unit area of land is much less than if the crops were directly consumed by people rather than animals.

Another type of food chain—known as a decomposer chain—involves the breakdown of dead organic matter, which includes plant and animal matter from all parts of the food chain. This breakdown is carried out by diverse organisms found in the soil or in water: fungi and bacteria are the main components of this group. The organic materials are decomposed to become simple inorganic compounds and the energy that began the process at the producer stage is dispersed as heat (see Fig. 1-16). These compounds, such as nitrogen and phosphorous, are then available to be used again in the development of plants.

Trophic Levels		Energy Flow	Respiratory Loss kJ/(m² a)	Decomposers	Species - Examples
Plants	Producers	87 100	50 130	1 695	Algae. Aquatic Plants
Herbivores	Primary Consumers	35 275	27 160	6 510	Small Fish Insects Snails
Carnivores	Secondary Consumers	1 600	1 320	190	Mice Birds
Top Carnivores	Tertiary Consumers	90	50	30	Large birds (owls, hawks)

Fig. 1-17. A small salt-water marsh ecosystem in the southern USA showing energy flow through the trophic levels in a period of one year. Based on Odum, H.T., 'Tropic Structure and Productivity of Silver Springs, Florida'. *Ecological Monograph*, V. 27, pp.55-112, 1957.

Of course the whole process is much more complicated than the above description indicates. One herbivore may consume many different species of plants and in turn be consumed by more than one species of carnivore. Thus food chains are linked together forming *food webs* and these may be very large and complex.

Since the chains and webs proceed from very small organisms to progressively larger ones—that is, a large number of organisms at one trophic level are necessary to support one larger animal at a higher level—the different trophic levels form a series of pyramids. One such pyramid for a very small ecosystem is shown as Fig. 1-17. In this pyramid each level represents the total flow of energy used at that level in the development of new plant or animal tissue as well as that lost through respiration and decomposition. Such information is very difficult to determine but it is extremely important as it shows how energy is captured and passed on from one trophic level to the next. Thus solar energy is transformed by photosynthesis into plant material in the example shown in Fig. 1-17 to the extent of 87 100 kJ/(m² a). Of this, 35 275 kJ/(m² a) are passed on to consumer organisms and the rest is lost to respiration and decomposition.

STUDY 1-8

1 Explain with the help of Figs. 1-16 and 1-17 the following characterstics of ecosystems.

a Different ecosystems have different species yet all require three basic biological elements: producers, consumers, and decomposers.

b Energy flow in an ecosystem is lost along the food chains and flows in one direction only.

c The various components of an ecosystem are highly interdependent.

2 In the past people have sought to achieve control over nature. Pursuing this objective has altered the functioning of most of the earth's ecosystems. Some of the important ways in which this has occurred include the

remodeling of landscapes using fire, axes, ploughs, construction, and chemicals. Furthermore the earth's ecosystems have been affected by the introduction of new animal and plant species, particularly through domestication. In addition new plant and animal pests (for example, malaria mosquitos) as well as organisms that cause disease (Dutch elm disease) have arisen.

a Attempt to produce a simple description of an ecosystem that might have existed in your community before the area was settled.

b Using the material above comment on the changes that have occurred. Which changes do you feel were good ones and which had a harmful effect?

c Why are we more concerned about our environment and its ecosystems today than our ancestors were 50 or 100 years ago?

d 'It appears healthier for man to regard the planet less as a set of commodities for use and more as a community of which he forms a part.' Explain and discuss.

e Give some examples of natural events that have also changed ecosystems.

2/ THE EARTH'S WATERS

We often forget that the surface of the earth supports an immense volume of water. The water on the earth together with the water in the atmosphere surrounding the earth is referred to collectively as the *hydrosphere*. Perhaps many of us also forget or take for granted the importance of water in all the earth's ecosystems and as a necessity of life itself. Thus a study of physical geography must examine some of the more important aspects of water on the earth.

Approximately 70 per cent of the earth's surface is covered by oceans and seas. These waters are deep, their volume being much greater than that of the land that lies above sea level. In addition, fresh water (water relatively low in dissolved salts) in the lakes and rivers on the land covers another 5 per cent of the earth's surface. Fresh water in frozen form on Antarctica and Greenland covers another 3 per cent. As well as occupying large areas of the earth's surface, water is also found in the air in vapour form and in the soils and rocks of the earth's crust in small but significant amounts.

Water is an important and necessary ingredient of all living things. The human body is over 60 per cent water, and water is our most basic need (about 2 L per person per day); people die of thirst long before they starve to death. We also require water for many other purposes. Some of the more important include washing, cooking, sewage disposal, agriculture, manufacturing processes, cooling, transportation, the production of electricity, and recreation.

Plants also require water to grow. For example, the growth of 1 kg of dry wheat requires about 475 L of water while the same quantity of rice takes approximately 1600 L. The most important characteristics of water are its abilities to dissolve other substances and to act as a medium of transportation. Thus the compounds vital to plant and animal life are transported in solution within living things. The water itself (an inert solvent) undergoes no chemical change, and when its function as a medium of transport is completed, the water is purified by one of several means (eg. evaporation), and is ready for re-use.

Water is also an important element in shaping the surface of the earth's crust. Water enters the cracks and crevices of the crust and, in various ways (such as freezing and thawing), causes the outer layer of rock to break up. Rivers and streams then carry these broken particles, usually from the higher areas of land, and deposit them in lower areas. While this process of land sculpturing is more complicated than the above suggests (it is discussed more fully in Chapter

8), many of the surface characteristics of the crust are produced in this way.

Finally, though the amount of water as vapour in the atmosphere is small, its influence on the biosphere is vital. Water vapour is the source of all precipitation and also influences temperatures at the earth's surface. This subject is more fully examined in Chapter 3.

CHANGES OF STATE

Unlike most of the substances found in the biosphere, water occurs in all three physical states—liquid, solid, and gas. It is constantly changing from one state to another in response to changing temperatures and pressure conditions. These changes are so balanced, however, that the amount of water in each state at any time remains almost constant.

The change in state from liquid to vapour is called *evaporation*. Although this may occur from any surface where water is present, most of the water vapour in the atmosphere has evaporated from the oceans. The process of evaporation requires considerable amounts of energy. For example it takes more than eight times as much energy to evaporate a gram of water at 0°C as it takes to convert a gram of ice into water. Through the process of evaporation, heat in a potential or latent form is transferred from the surface of the earth to the atmosphere. The importance of this phenomenon as a means of heating the atmosphere is examined in Chapter 3.

The amount and rapidity of evaporation from a water surface depend on a variety of factors. These include the temperature of the air, its aridity, and its movement, as well as the water's temperature and the area of its surface exposed to the atmosphere. Evaporation is at a maximum in the low latitudes because of the greater heat; it decreases polewards.

The change of state from vapour to liquid is called *condensation*. This occurs when moist air is cooled, thus reducing its capacity to hold water vapour. Water droplets resulting from condensation may remain suspended in the air as clouds or fog. When they combine or otherwise become too large to remain suspended, they fall as precipitation.

Sublimation is the change of state from a solid to vapour. When dry air with a temperature below 0°C comes in contact with snow or ice, some particles of snow or ice will pass directly into the vapour state. The opposite process—where vapour is changed to ice crystals—is also called sublimation. The formation of frost is an example of this.

THE HYDROLOGIC CYCLE

A small but important quantity of the earth's water is constantly moving in a vast series of pathways that are collectively referred to as the *hydrologic cycle* (Fig. 2-1). This cycle can be thought of as including a series of storage areas interconnected by transfer processes. Most of these processes are driven by solar energy. They include evaporation, the transportation of water vapour by wind, condensation, precipitation, and runoff (the transfer of water across the land surface). Figure 2-1 is a simple model illustrating the quantity of some of these flows within the hydrologic cycle.

At any time most of the earth's water is held in storage—97 per cent in the oceans and 2 per cent in glaciers. The remaining 1 per cent is the active component in the hydrologic cycle. This water moves fairly rapidly between the oceans, atmosphere, and land surface. For example, it is estimated that at any given moment the atmosphere contains only about ten days' supply of rainfall. It should also be emphasized that this 1 per cent of the earth's water makes life on land possible.

Figure 2-1 demonstrates that water falling as precipitation on land may follow one of three courses. Some may be evaporated or

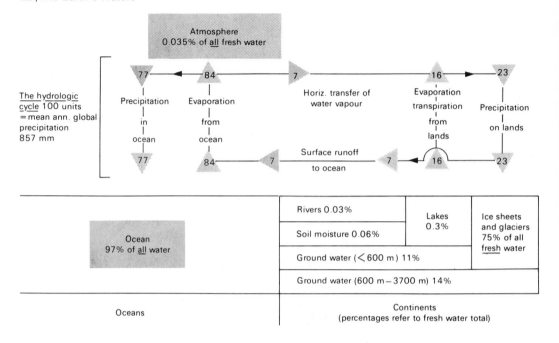

Fig. 2-1. The hydrologic cycle. A simple model showing the flows and storage within the hydrologic cycle. The mean annual global precipitation of 857 mm has been reduced to 100 mm so that the different values in the hydrologic cycle represent percentages. Of the total water supply on the earth of 1.356×10^9 km^3, 1.3×10^9 km^3 are in the oceans, 2.9×10^7 km^3 in the glaciers and ice caps, 1.3×10^4 km^3 in the atmosphere, and 8.6×10^6 km^3 on the land. This last figure can be broken down as follows: 1.2×10^3 km^3 in rivers and streams; 1.2×10^5 km^3 in fresh-water lakes; 1.04×10^5 in salt lakes and in land seas; 6.6×10^4 as soil moisture and seepage; and 8.3×10^6 as ground water, i.e. below the water table.

transpired back into the atmosphere; some runs off by way of rivers to the oceans (some of this runoff may be temporarily stored in lakes or ponds); and the remainder sinks into the earth. The latter is referred to as *ground water*. The three courses are closely interlinked. For example, evaporated water may be re-precipitated to become surface water elsewhere, while some of the surface water in streams and lakes may eventually sink into the earth and become part of the ground water system. Ground water flows beneath the surface and much will reappear, for example as springs, to be temporarily stored in lakes or to become part of the surface runoff.

It is estimated that water spends between ten and one hundred days on land unless it becomes part of the ground water system. Ground water is usually held for a much longer period. In fact there is evidence in some places that ground water has been trapped deep beneath the surface for thousands of years.

THE CHARACTERISTICS OF WATER

Like air, water is an indispensable resource to all forms of life. None of its uses, however, alters the quantity of water present in the biosphere. The same amount is available today as existed millions of years ago. Nor is there reason to doubt that the same amount will not continue to be available thousands of years from now. However, as we shall see later in this chapter, the quantity of water—particularly fresh water—is unevenly distributed over the earth's land surface. This results in serious shortages in some areas. In addition, while the total quantity of water on earth is fixed, its quality is not. The pollution of both fresh and salt water is a problem of world-wide proportion. In order to understand these and other problems, the following characteristics of water must be understood.

1 Water is constantly moving. Rivers drain the land through an intricate network of channels. Ground water moves more slowly but much of the water that sinks into the ground eventually re-emerges in streams, lakes, or the oceans. The oceans also move in vast surface and sub-surface currents. All these motions are interconnected in the hydrologic cycle.

2 Water, as we have seen, is used for a vast number of purposes. It is the universal resource, the medium on which all life depends.

3 More than any other resource, the uses of water are interdependent. This characteristic causes problems, for many of the uses of water are not compatible. For example, the use of a river or lake for drinking water or swimming conflicts with its function as a receptacle for waste disposal. The use of a river for transportation makes it difficult to use the same river for the generation of hydro-electric power.

4 Most waters cannot be privately owned, and consequently their management is the responsibility of governments. However, surface and ground water often make this a difficult task by flowing from one political jurisdiction to another. For example, the establishment of standards of water quality in the Great Lakes requires co-operation not only between the governments of Canada and the United States but also between the province of Ontario and eight states of the U.S.A.

Management of the waters of the oceans is even more complicated because of the inability of nations to agree on questions of political jurisdiction. This is discussed later in the chapter.

STUDY 2-1

1 For the earth as a whole, water retention on the land surfaces is greatest in March-April. It is estimated that by September-October the oceans are 1 to 2 cm higher than in the earlier period and contain 7.5×10^{18} cm^3 more water. Why is this so?

2 Figure 2-2 shows the major storage areas and some of the transfer processes in the hydrologic cycle. Arrows indicating the various pathways have been left off. Using a photostat or sketch of this figure, mark on the arrows to show the pathways and briefly note the nature of the different transfer processes.

3 How can the hydrologic cycle be affected by human interference?

4 Most of the water required for human use is fresh water; at least 97 per cent of the water on earth is salt water. Referring to the hydrologic cycle, explain how fresh water is produced, that is, where and how does the distillation process occur?

5 The shape of the continents is largely influenced by sea level. Using a map showing elevations below sea level as well as above it, draw two maps of North America. On the first one, show what the continent would look like if ocean levels fell by 100 m, and on the other show the continent as it would appear if levels rose by 100 m. What might cause ocean levels to change?

WATER ON THE LAND

Most of the water on the land surface is fresh water, whether in lakes, rivers, or ground water. In the previous sections we have discussed how this water arrived on the land and stressed the importance of its role in sus-

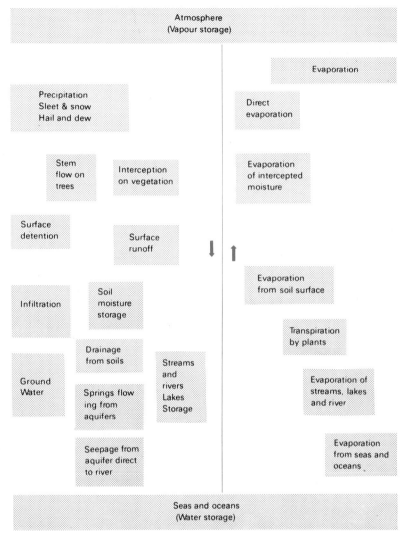

Atmosphere
(Vapour storage)

Evaporation

Precipitation
Sleet & snow
Hail and dew

Direct
evaporation

Stem
flow on
trees

Interception
on vegetation

Evaporation
of intercepted
moisture

Surface
detention

Surface
runoff

Evaporation
from soil surface

Infiltration

Soil
moisture
storage

Transpiration
by plants

Drainage
from soils

Streams
and
rivers
Lakes
Storage

Ground
Water

Springs flow
ing from
aquifers

Evaporation of
streams, lakes
and river

Seepage from
aquifer direct
to river

Evaporation
from seas and
oceans

Seas and oceans
(Water storage)

Fig. 2-2. A schematic and simplified diagram of the hydrologic cycle. Storage areas and transfer processes are indicated but arrows representing transfers have been omitted.

taining life. In subsequent chapters we will examine the importance of water in the atmosphere, water as a factor in the creation of landforms, and the role of water in the development of soil and vegetation.

Lakes

Lakes occur wherever there is a natural basin with a restricted outlet and where there is sufficient input of water to keep the basin filled. These conditions are not found on all parts of the land and consequently lakes are unevenly distributed. For example, the northern parts of North America and Europe have many lakes, but there are very few in South America, Africa, and Australia. The presence or absence of lakes is entirely explained by climate and whether or not basins with restricted outlets are formed and, once formed, remain in existence. The best examples of surfaces with a large number of lakes are areas of humid climate with hard rock that has recently been glaciated, such as the Canadian or Baltic Shields.

While the proportion of the total water of the earth contained in lakes is very small, lakes perform many useful functions where they exist. They serve as natural reservoirs storing water during wet periods, thus helping to maintain water levels in out-flowing rivers during dry periods. This storage function also protects areas downstream from destructive floods and erosion when periods of high water occur, for example during the spring breakup. Lakes also provide water storage for irrigation purposes. Large lakes in particular affect the climate near their shores by moderating temperatures and providing a source of moisture for precipitation. Some, such as the Great Lakes, are important transportation routes. Many dams have been constructed, creating man-made lakes to provide a regulated flow and 'head' of water for the generation of hydro-electric power. Lakes in remote regions provide part or all of the habitat necessary for many animals, birds, and fish. They also provide water for cities and towns, and then receive the same water in return as sewage. Lakes perform important recreation functions; those located near centres of population are often popular refuges for harassed urban dwellers.

Rivers

Rivers are essentially the surface flow (also called *runoff*) of excess water from precipitation. As this water seeks to escape to lower levels and eventually to the sea, it carves out drainage systems. Each drainage system consists of a series of converging streams in which excess runoff is collected from almost every square centimetre of land in the area

and carried into progressively larger streams and sometimes lakes, until it enters the ocean or evaporates. Not all rivers become larger as they proceed downstream. In desert areas, where evaporation rates exceed precipitation, the stream flow in a river may actually decrease downstream.

The term 'river' and 'stream' are used to denote not only the flow of water in channels but also the channels in which water flows. Streams that contain water only during or immediately after a rainstorm are called *ephemeral*; those that carry water for a longer period—often several months—are referred to as *intermittent*; and streams that contain water at all times are called *perennial*. In general the three types are related to the depth of the water table (explained on p. 27). Permanent stream flow is possible only where the water table is high enough to provide a continuous flow of ground water into the stream all year round.

A map of a drainage system in a humid part of the world is similar in some ways to the branching pattern of a deciduous tree. The trunk, of course, is the master stream, with branches joining it right down to ground level. At higher levels the trunk gradually divides into large branches that, in turn, re-divide many times into progressively smaller branches. The term *dendritic* is used to describe the tree-like pattern. In general a dendritic pattern will usually develop where the underlying rock is uniformly resistant to erosion. Where rock types with different resistance occur, other patterns or river systems with no discernible pattern (Fig. 2-3) will emerge.

In humid regions there are no gaps between drainage basins. Although the basins seldom intermesh as the trees in a forest may, the outermost branches of one system may almost meet the outermost branches of an adjoining system. Although drainage basins exist in arid or sub-humid regions, they may be more difficult to recog-

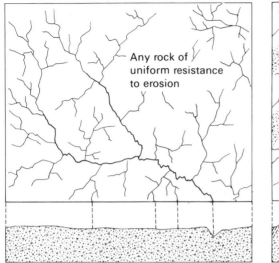
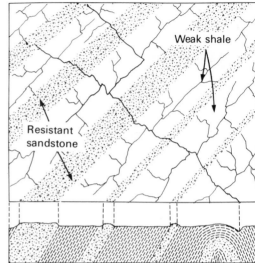

Fig. 2-3. The effect of rock structure on stream patterns.

nize. Instead of draining to the sea, runoff in arid regions often flows into depressions where it evaporates or sinks into the ground. In the larger depressions the runoff forms a lake, the level of which is controlled through evaporation.

STUDY 2-2

1 Figure 2-4 is a reasonably detailed map of the river systems of France. Using a sheet of tracing paper, draw the main rivers shown on the map and indicate by colour each of the drainage basins. Put an arrow alongside each of the main rivers to indicate the direction of flow. What can be deduced about the surface characteristics of France from an analysis of this map?

2 Referring to Fig. 2-4 and an atlas, mention the general characteristics of drainage basins, indicating how the basins may vary in size, shape, elevation, and the patterns of their rivers and streams. Illustrate these characteristics and the terms listed in Question 3 on two general cross-section sketches—one from the edge of the drainage basin to the mouth of the river and the second across the middle of the drainage basin at right angles to the first.

3 Define the following terms: divide (also called watershed), interfluve, tributary, confluence, main river, right bank, left bank, mouth, and source. Make a sketch of a drainage basin and locate an example of each on it.

4 On Fig. 2-4 the small square south of Paris represents an area of 6500 km². Enlarge this square to a size of 25 cm² (5 cm × 5 cm), drawing in the rivers that appear on the original map. What is the scale of the new map? This scale makes it obvious that the river system as shown is incomplete. Draw in additional rivers and streams in the most likely locations.

Groundwater

Water held below the surface within the weathered rock materials (regolith) or in the spaces that occur within the bedrock itself is referred to as *ground water*. The place of ground water in the hydrologic cycle has

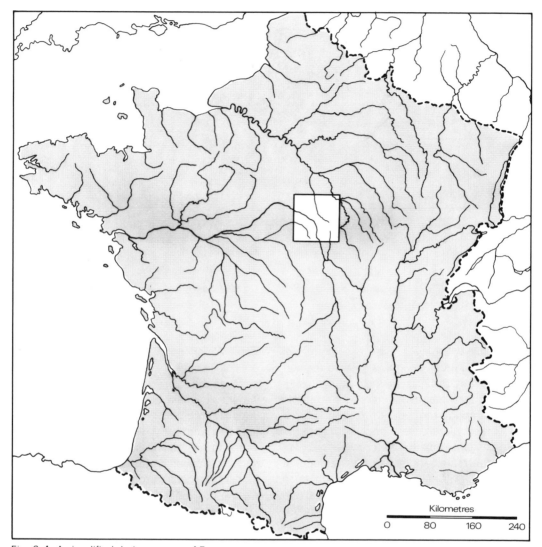

Fig. 2-4. A simplified drainage map of France.

already been discussed on page 22 and shown in Fig. 2-1. Figure 2-5 shows the three principal elements in the occurrence of water in the area immediately below the surface: the *zone of aeration* where both air and water exist in the pore space between rock and soil particles; the *zone of saturation* where the pore spaces are completely saturated with water; and the *water table* itself, which is the top of the zone of saturation. Another element of considerable importance is the *zone of soil moisture* immediately below the surface. It normally contains more moisture than the rest of the zone of aeration.

The depth of the water table varies. In most humid areas it is only a few metres or tens of metres below the surface, while in arid areas it may be hundreds of metres

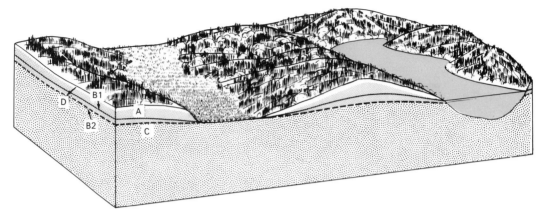

Fig. 2-5. A represents the zone of aeration, B^1 and B^2 the water table (i.e. the top of the zone of saturation), and C the zone of saturation. D is the zone of soil moisture. Why does the level of the water table fluctuate from position 1 to position 2? What would control the depth of the zone of saturation?

underground. The water table is seldom level but follows a profile roughly similar to the surface above it because of the natural resistance that the water encounters as it attempts to seep through the ground. In other words, the water table will develop a configuration with sufficient slope for water to seep downhill and eventually intersect the surface at lakes, streams, and springs. Through such seepage a balance is achieved between the amount entering the ground and the amount being discharged. Except for short-run or man-induced variations, the level of the water table will be relatively stable.

Ground water may also be found at considerable depths below the surface and under different circumstances than those shown in Fig. 2-5. Where permeable layers of rock such as sandstone (called aquifers) alternate with rocks of low permeability such as shale (called aquiclude), conditions may be favourable for *artesian wells* or *artesian springs*. As illustrated in Fig. 2-6, these rock layers must be gently sloping so that the higher end of the rock formation is exposed to precipita-

tion. Water permeates down through the aquifer from this catchment area and because the rock strata is gently sloping, this water will usually be under pressure. If a well is drilled through the aquiclude into the aquifer, this pressure will normally cause the water to rise to the surface. An example of such an aquifer is the layer of Dakota sandstone that dips from the Rocky Mountains eastwards under the interior plains of the United States. Wells drilled into this aquifer have been an important source of water for these areas since they were first settled.

STUDY 2-3

1 Figure 2-5 shows that the level of the water table varies from place to place and fluctuates seasonally. It is also affected by the number of wells that are drawing water from it. Explain why these changes in the level of the water table occur, and make a diagram to indicate how these factors would affect well construction.

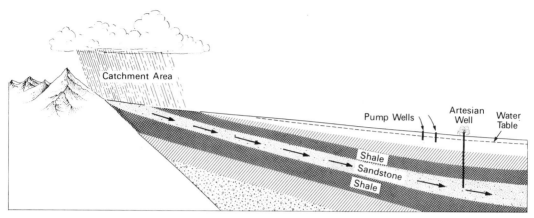

Fig. 2-6. Ground water held in a sandstone aquifer in gently sloping sedimentary rock.

WATER SHORTAGES

Water shortages or insufficiencies can take different forms in different places. In areas around large and growing cities shortages of fresh water occur as a result of the ever-increasing demands on existing water supplies by growing populations and large industries. Water shortages also occur as a result of periods of below-average precipitation lasting months or years. While most prevalent in arid regions, these dry periods can affect humid areas, i.e. the severe periods that occurred in the British Isles and parts of western Europe in 1976 and in parts of western North America in 1976 and 1977. The more urgent water shortages, however, are located in the arid regions of the world. The following material deals specifically with the most serious of these—the creeping spread of the world's deserts.

In the late summer of 1977 a United Nations conference convened in Nairobi, Kenya. Its purpose was to consider the expansion of deserts, particularly in the Sahel region of Africa (Fig. 2-7), the area of land along the southern flank of the western part of the Sahara Desert. This growth of deserts is referred to as *desertification*. It is responsible for the destruction or deterioration of agriculturally usable land at rates estimated for the entire world as high as six million hectares a year. In general terms the task of this UN conference was to find ways to slow down or even reverse the process of desertification. The results of their studies are critical to the future of these regions in view of their serious food shortages in past years and the death of large numbers of people from starvation. (While these conditions affected a number of areas, they were particularly severe in the Sahel during the drought period between 1968 and 1972.) Considering the growth of the world's population, and the increasing need for space and food, the problem has global implications.

The causes of desertification are complex and incompletely understood. In the first place, desert regions normally experience great variations in precipitation. Some

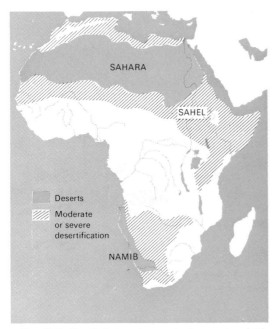

Fig. 2-7. The Sahel region and other parts of Africa experiencing some form of desertification. The Sahel refers to parts of Mauritania, Senegal, Mali, Upper Volta, Niger, and Chad. The drought conditions in the early 1970s were equally as severe to the east in Ethiopia, Sudan, Somalia, Kenya, and Tanzania.

authorities claim that the below-average precipitation in recent years is merely a repetition of similar circumstances that have been reported at various times since the beginning of recorded history. Indeed the resilient nature of desert ecosystems is proof of this phenomenon. After months or even years of drought, when the desert landscape seems devoid of all forms of life, a short period of heavy rain brings about a rapid reappearance of plants and animals.

Desert peoples have also developed fairly sophisticated techniques for survival. The most obvious of these is nomadism. During dry periods people simply move into areas where conditions are not as difficult. Traditionally such nomadic people have used live-stock as a source of food, transportation, hides for clothing, and so on. It has been customary to keep different kinds of livestock because each can survive on a different set of plant species commonly found in the same general area. Some desert peoples reduce their risks further by planting a few hectares in crops when conditions are suitable. Desert people have also been noted in the past for their low rate of population growth, and the close association between families and tribes was also an important factor enabling the people to live through difficult times. In recent years, however, there have been many changes. Populations in and around desert areas have grown considerably. Many formerly nomadic people have moved into permanent settlements. Furthermore, innovations such as deep wells have lulled many people living in dry regions into believing that they are no longer as dependent on rainfall as in the past. Herds have been enlarged and more land has been planted with crops such as cotton and peanuts. Such land use reflects a short-term view of the desert environment. When the recent periods of drought began it was soon apparent that the land had been overused. Unfortunately the traditional response of moving away was no longer practicable. As a consequence many people died and many were forced to leave the stricken regions as refugees.

The recent changes in the patterns of desert life suggest that desertification—while accelerated during times of drought—

Fig. 2-8. An oasis on the desert fringe. Given adequate water the soil in such areas can be quite fertile.

Fig. 2-9. The sparse vegetation on the fringes of the Sahara desert in the Sudan has been damaged by overgrazing and overcultivation. Patches of bare sand link up into mobile dunes that can be blown by the wind to encroach on farmland, rangeland, or oases.

may in fact be more a result of human pressure on the land through overcropping and overgrazing. Scientists who contend that human misuse of the land is the main cause of desertification point to the fact that the edge of the desert is not advancing on a broad front, as one might expect if precipitation was the only factor involved. Instead desertification is occurring in various areas that do not border the desert itself. In such areas, damage to the vegetation, soils, and other components of the ecosystem is a result of one or more of the following factors: the overuse of the soil for crops; the effects of cattle trampling and overgrazing vegetation; and the destruction of vegetation as a result of overcutting for firewood. There is little doubt that these practices are occurring and are causing great damage. It is not clear whether they alone are sufficient to cause desertification. In other words, is the damage caused by people permanent or does it merely add to the damage caused by years of low precipitation? Furthermore, there is strong evidence to indicate that poor land-use practices can also contribute to climatic change. Thus drought feeds on drought. (This is referred to in the section on climatic change on p. 114).

Finally, we do not yet know whether long-term climate change is the principal cause of desertification. Opponents of climatic change as an explanation refer to the relief of the Sahel drought with the return of normal precipitation in 1974, which seemed to reverse or stop desertification. However this does not necessarily refute the theory of climatic change. If the climates of desert regions are changing, such changes would not occur abruptly. It is unlikely that we would recognize the change until the whole process was well under way.

STUDY 2-4

1 Identify the water supply source for your community. Are you aware of any problems of water shortage in your community? If a shortage exists, how was it caused?

2 How are problems of water quantity and water quality sometimes related?

3 Why is it difficult to unravel cause and effect in the process of desertification?

4 What do you see as a sensible solution to the problems of desertification? Prepare a statement containing positive measures that you believe would be appropriate to help solve the water problems in a region such as the Sahel.

THE OCEANS

The importance of the oceans as inexhaustible (but not indestructible) reservoirs justifies their inclusion in a study of the earth's biosphere. The oceans on the earth collectively contain 97 per cent of the world's water and cover over 70 per cent of the earth's surface. They are, however, important for many other reasons. As reservoirs they store a large quantity of solar heat which, when released, is an important factor in moderating the earth's temperatures. The oceans are also the only source of the fresh water that falls on the land as precipitation. Within the oceans a vast array of plant and animal life forms exists —amazingly about 80 per cent of the living organisms on the earth are in the sea. Certain plants (plankton) provide as much as 70 per cent of the oxygen in the atmosphere, while many of the animals are a minor but still important supply of human food. Oceans also contribute to human life by providing minerals, a means of transportation, and even a place for recreation.

The Features of the Ocean Bed

The continents on which people live are large islands raised above the general level of the earth's crust. Edging these continents are those parts of the ocean most familiar to

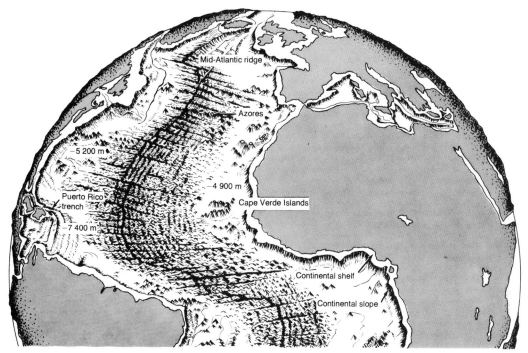

Fig. 2-10. The North Atlantic Basin as it would appear if all of its water were drained away. Among the features shown are the continental shelves on either side of the Atlantic. Although not entirely obvious from the diagram, these shelves fit together quite precisely, thereby supporting the theory of continental drift first proposed by Alfred Wegener in 1915. Another prominent feature is the Mid-Atlantic Ridge, a craggy undersea chain of mountains, 16000 km long, whose depth below sea level ranges between 540 and 1800 m. The ridge is split by a rift valley through which volcanic eruptions are constantly occurring. This ridge, as well as the east-west faults, is related to the spreading of the ocean basin. This spreading occurs as the sea-floor and the continents move apart at approximately 3 cm/yr (see p. 000).

man. Here the waters are shallow, for beneath them are gently sloping platforms known as *continental shelves* (Fig. 2-10). They belong, structurally, to the continents they surround rather than to the ocean floor. Indeed, at various times in the past, many of the shelves have been above sea level. Such shelves are generally quite smooth with sloping floors that slant away from the continents; for the most part they are less than 200 m deep. Their width varies greatly, from over 300 km off the Atlantic provinces of Canada to as little as a few kilometres off much of the coast of California.

Continental shelves end in very steep slopes known as *continental slopes*. These exhibit some of the greatest changes in elevation to be found on earth. Many plunge to depths of 1500 to 3000 m at gradients of about one in five. At the bottom the slope lessens rapidly in a zone known as the *continental rise*, which descends to depths of 5000 m.

The *deep-ocean floor* begins at the bottom of the continental rise. It underlies the greatest part of any ocean, generally at a depth of between 4500 to 5500 m. Much of the deep-ocean floor, called the *abyssal plain*, is unusually flat. This is thought to be the result of deposition spread out over long periods of time in thin sheets that bury most of the irregularities in the ocean floor. There are still, however, significant irregularities on the deep-ocean floor. The most notable features are the highly fractured mountain ranges or ridges beneath the Atlantic, Pacific, and Indian Oceans (these ridges are discussed in conjunction with 'Sea-Floor Spreading' in Chapter 7). Some of these ranges and certain isolated volcanic mountains extend above the surface of the water, forming countless islands of different sizes, particularly in the southwest Pacific. There are also very deep, steep-sided *trenches* below the normal depth of the deep-ocean floor. They are mostly adjacent to the continental slopes and often occur next to mountainous *island arcs* such as the Japanese, Kurile, and Indonesian Islands. They reach their maximum depth in the Marianas Trench (11 022 m). The origin of these trenches is explained in Chapter 7.

Ocean volcanoes, known as seamounts, are one of the most interesting features of the deep-ocean floor. Their isolated peaks rise 900 m or more above the ocean floor. There are estimated to be several hundred in the Pacific alone. While most are below sea level, some form islands that in certain locations have been ringed by coral reefs. One particular type of island is the coral atoll, which was formed when the seamount submerged while the coral continued to grow. This resulted in a doughnut-shaped coral island encircling a lagoon whose waters cover the submerged volcano. One of the more famous of these is Bikini Atoll in the South Pacific, which was used as an atomic testing ground by the United States after the Second World War.

STUDY 2-5

1 The sketch of the floor of the North Atlantic Ocean (Fig. 2-10) illustrates many of the features just described. Referring to the diagram and to an atlas, draw a rough east-west cross section of the basin of the North Atlantic Ocean, showing the adjoining land areas and labelling the various features already mentioned.

2 List and explain the various ways in which the oceans are important to life on the earth.

Sea Water

Sea water is a salt solution or a brine. About two-thirds of its salt content consists of sodium chloride (common table salt). Most of the remaining third includes such salts as magnesium chloride, sodium sulphate, calcium chloride, and potassium chloride. Trace quantities of at least half the elements known to exist on earth—including all the gases of the atmosphere—are also included. It has been generally believed that the salts presently found in the oceans came from soluble minerals carried from the land by rivers and ground water to the sea. Since evaporation from the oceans leaves such minerals behind, they have continued to concentrate in the oceans. Although the quantity of salts dissolved in fresh water is very small, the present level of salinity in the oceans has probably been established by the addition of these small quantities over the history of the earth. The difficulty with this explanation is that there is good evidence that the salinity of the ocean has not changed for at least the past 200 million years. It is probably true that the real explanation of the ocean's salt content is not totally understood. Some scientists believe that a series of complex cycles are responsible. In these cycles salts are both removed and replenished in the ocean by such mechanisms as sea-floor spreading (which causes the release of new or juvenile water from the earth's interior), volcanic eruptions, precipitation, and rivers. Over the long run losses and gains are balanced.

The degree of salt concentration is referred to as salinity. It varies from place to place depending on a number of factors. The most important is the relative rate of precipitation and evaporation. Heavy rainfall obviously lowers the salinity by dilution, while evaporation removes water and causes greater concentration of salts. The degree of concentration is also influenced by ocean currents, the outflow of large rivers, and the presence of melting ice. Average salinity is approximately 35 parts per 1000, an amount that has been recorded in many parts of the North Atlantic. Generally the salinity is average in equatorial areas, increases in the sub-tropical oceans and seas, and then decreases polewards, although there is considerable variation from place to place. For reasons not fully understood the Pacific Ocean is less saline than the Atlantic. Even greater variations occur in partly enclosed bodies of water. In the Baltic Sea, for example, salinity is around 7 parts per 1000, while in the Red Sea it is around 40 parts per 1000. This high concentration occurs because the Red Sea is almost entirely enclosed by land. Furthermore, the Red Sea experiences a desert climate. The most saline waters on the earth are found in inland seas such as the Caspian, Aral, and Dead Seas. The water levels of these seas have dropped so low that none of them flows into rivers. This situation has occurred because the rate of evaporation from their surfaces far exceeds precipitation and the inflow of water from rivers and streams into the seas. Consequently they accumulate dissolved salts but do not transfer them elsewhere.

The surface temperatures of sea water vary from a high of 26°C in the tropics to -2°C (the freezing point of salt water) at the poles. Although the temperature decrease in a north-south direction is far more regular over the oceans than over land, there are variations caused by the influence of air coming from the land and also by ocean currents. One particularly noticeable effect of the movement of air occurs during winter in the middle latitudes on the eastern sides of both Asia and North America. Here cold air coming off the land has a marked cooling effect on the ocean. However, the movement of water in ocean currents is of greater impor-

tance in causing variations. In some places cold water moves into the lower latitudes while in other places warm water moves into the higher latitudes. Thus the average sea temperature off the coast of Labrador is nearly 10°C colder than water in the same latitude off the coasts of the British Isles. An examination of Fig. 2-13 will reveal how different ocean currents are responsible for these variations.

Compared to variations in land temperatures, seasonal temperature variations in the surface layers of the oceans are very small. They range from a mere 1-3°C in the tropics to only 6-7°C in the middle latitudes. Daily changes seldom exceed more than one degree between day and night. The causes and consequences of these small temperature variations are discussed in Chapter 3 (p. 52).

Waves, Tides, and Currents

Waves are created and set in motion by the friction produced between moving air and the surface of a body of water. The size of the waves depends on the velocity and duration of the wind and the extent of water over which the wind can blow freely (fetch). As waves become larger, it is easier for wind to pile up the water even further. Under optimum conditions, waves 12 to 15 m high may be produced.

Until a wave actually breaks along a shoreline, the particles of water involved in the wave motion are engaged principally in a vertical rather than a horizontal movement. In other words, the water of the wave does not move across the ocean, only the form of the wave moves. As shown in Fig. 2-11, when a wave passes a given point each water particle traces a circle whose diameter is equal to the height of the wave. As a wave approaches a shoreline, however, it undergoes a change. The water molecules are slowed down at the bottom of their circle by contact with sand,

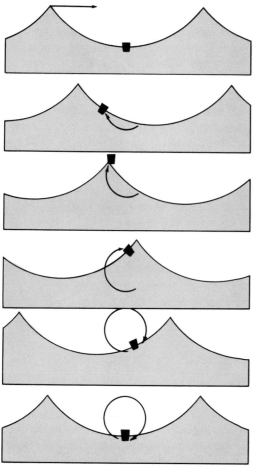

Fig. 2-11. The float in this diagram demonstrates that while a wave travels forward, the water itself does not move. The float, which can be thought of as a water molecule, rotates in a circle as the wave passes, returning to within a centimetre or two of its original position after the wave has passed. Water molecules below the surface complete a similar circle, although the deeper the molecule the smaller its orbit.

rock, or gravel, while the crest moves ahead and breaks along the shore. Waves nearing land range themselves in lines more or less parallel to the shore, no matter what their original direction may have been. As a result, breaking waves have a tendency to wear away and at the same time straighten irregular shorelines.

On occasion, waves known as *seismic* waves or *tsunamis* are sent across the ocean by a violent disturbance on the ocean bottom or along the coast. They are usually caused by earthquakes or volcanic activity. Tsunamis are quite different from the waves built up by wind; they measure 100 km or more between crests and travel at speeds up to 800 km/h. The speed of the tsunami wave is proportional to ocean depth. This means that the wave form travels at variable speeds, reaching a maximum over the deepest water. Since the wave length of a tsunami is so great, in the open sea it seldom rises more than a metre and is not likely to be noticed by passing ships. However, when a tsunami approaches land its wavelength and speed decrease and its height (amplitude) increases. At this stage the tsunami becomes a terrifying natural phenomenon. In April 1946 four such waves, caused by an earthquake off the coast of Alaska, struck Hilo harbour on the island of Hawaii. These waves, almost 15 m high, smashed ships, destroyed buildings along the coast, washed away highways, and killed 160 people. As a consequence of this and other similar disasters, a detection and warning system (the Tsunamis Warning System) now exists. It has effectively reduced casualty figures.

Tide is the name given to the periodic rise and fall of the sea. Tides are mainly caused by the gravitational attraction of the moon and the sun pulling on the waters of the earth and causing them to bulge. The moon has a greater effect in the production of tides because it is closer to the earth. In relatively simple terms the gravitational pull of the moon causes two bulges in the earth's oceans (Fig. 2-12). One occurs on the side facing the moon and another, compensatory bulge occurs on the opposite side. As the earth rotates in an eastward direction these bulges move westwards around the earth. Since the earth makes one complete rotation every twenty-four hours, every place theoretically receives two high tides and two low tides each day. The actual interval from high tide to high tide is approximately 12.5 h, however small variations exist from one locality to another.

The difference in water level between high and low tide is known as the *tidal range*. In the open ocean the tidal range is a metre or less; on some exposed coasts it is commonly between 2 m and 4 m; in enclosed seas, such as the Caribbean, it is usually less than a metre. Owing to the funnel-like nature of the bay or estuary on which they are situated, certain localities experience a very great tidal range. For instance, the Bay of Fundy between Nova Scotia and New Brunswick has a normal tidal range between 12 m and 15 m.

During any single month the variation in tidal range in the same place is attributable to the moon's position in relation to the earth and sun. When the moon, earth, and sun are aligned in space (see Fig. 2-12), the gravitational pull of the sun augments the pull of the moon. This situation occurs twice a month during new and full moon phases. At these times high tides are higher, low tides are lower, and the tidal range over the 12.5 h interval is greater. Such a tidal extreme is referred to as a *spring tide*. When the moon, earth, and sun form a right angle in space (see Fig. 2-12), the gravitational pull of the sun at right angles to the moon slightly reduces the effect of the lunar pull. This situation occurs twice a month during the first and last quarter phases of the moon. As a result, high tides are lower than normal, low tides are higher than normal, and the tidal range over the 12.5 h interval is smaller. Such a tidal pattern is referred to as a *neap tide*. Yearly variations in the ranges of spring and neap tides are caused by variations in the distances between the moon, sun, and earth at the time of the tide. Tidal ranges are the greatest when tides occur during a full moon and also when the sun and moon are closest to the earth.

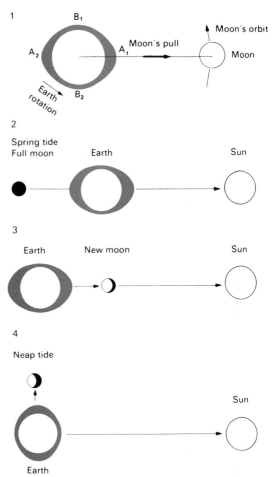

Fig. 2-12. In 1 the water at A_1 is pulled towards the moon. This causes water to be drawn from B_1 and B_2, resulting in low tides on this part of the earth. In 2, when the sun and the earth pull together as they do at Full Moon and New Moon, tides achieve their highest highs and lowest lows. The opposite situation is shown in part 4.

Tides are important in several ways. They have a significant effect on navigation in coastal waters and harbours and are a factor to be taken into account in the design of harbour installations. In bays and lagoons they are partly responsible for landforms called tidal or mud flats that are exposed at low water or spring tides. These vary greatly in

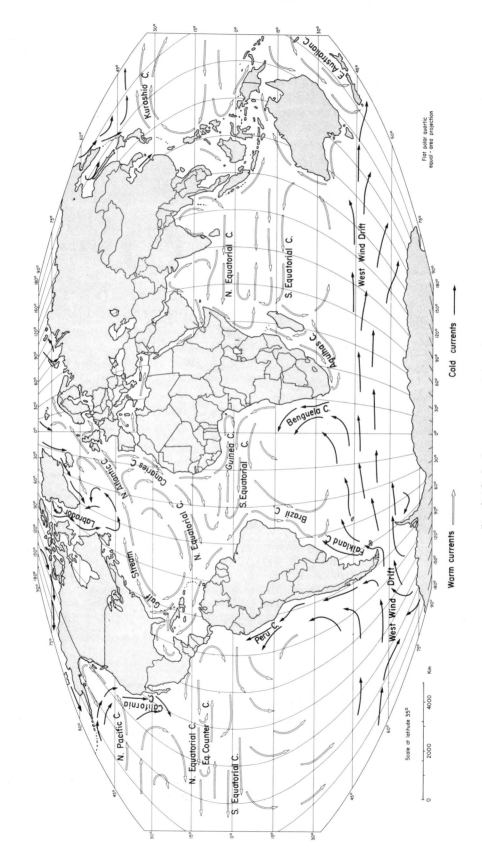

Fig. 2-13. Surface currents of the oceans.

width, and are an important part of the landscape where they form an exposed beach hundreds of metres wide at low water. Salt marshes sometimes develop on these flat surfaces and grow until the entire lagoon is filled except for a network of tidal channels. Tides may also be utilized in some areas with unusually high tides to provide hydro-electric power. Tidal-power plants are now in operation on the Rance River on the north Brittany coast of France and on the Barents Sea near Murmansk in the USSR. The development of this type of power facility has long been discussed for the Bay of Fundy (Minas Basin and Chignecto Bay), but high costs as well as engineering difficulties have prevented any scheme from getting under way to this point.

Currents

Though the waters of the oceans are involved in several types of movement that transport them over large distances, we are concerned here only with currents that operate at or near the surface of the ocean. Such surface currents are important for navigation and fishing, and as well have an indirect influence on the weather and climate of adjacent land masses.

Ocean currents are caused principally by the unequal heating of the surface waters by the sun. When cold water in polar regions sinks, it causes water less chilled to be displaced. This displaced water moves towards the equator and displaces the warmer waters of this region, which in turn move polewards. This pattern is complicated by several factors: the effect of the wind; the limits to water movement caused by the configuration of the continent's shorelines; and the effect of the earth's rotation.

One example of the influence of ocean currents on human activity involves the Humboldt (Peru) Current that flows off the west coast of South America. Normally the upwelling of cold water along the edge of this current provides nutrients to support a large population of anchovy. These fish, caught by Peruvian fishermen, have made this country one of the leading fishing nations of the world over the past few years. Occasionally, however, the Humboldt Current alters its course slightly, thus reducing the amount of upwelling. When this happens the anchovies disappear, causing great economic hardship to the Peruvians. Many other marine plants and animals also die, releasing large quantities of hydrogen sulphide into the atmosphere.

STUDY 2-6

The material below is arranged to lead to a general understanding of ocean currents, their patterns, and the factors responsible for the circulation of ocean water.

Draw a rectangle 10 cm × 15 cm, marking the longer side in degrees of latitude from pole to pole. Assuming the area within the rectangle to be the North and South Atlantic Oceans, complete the following:

a Referring to Fig. 2-13, draw on the rectangle the major currents of the North and South Atlantic, using single arrows to indicate direction.

b Name the major currents and indicate whether they are warm or cold ones.

c Although variations in temperature, salinity, and the configurations of ocean basins and coasts affect the movement of ocean water, the main cause of currents is the frictional drag of air moving over the surface of the water. Look ahead to Figs. 3-17 and 3-18 and describe the relationship between the movement of air and the generalized scheme of ocean currents shown in the rectangle. Very careful observation will indicate that the ocean currents shown in the rectangle move even farther to the right in the northern hemisphere and farther to the left in the southern hemisphere than the direction of the wind. What causes this?

d Can a similar generalized scheme for the movement of water in currents and drifts be applied to the Pacific Ocean? Does the same pattern emerge? Are there any significant variations?

e How does the configuration of the coast cause variations in the general pattern of currents in the North Atlantic and the North Pacific?

Life in the Ocean

In the sea, as on land, plants form the basis of all animal life and are referred to as the primary producers (see p. 17). There are two basic marine plant types: those that are fixed to the sea bottom in shallow water, such as seaweeds and kelps; and the free-floating *phytoplankton*. The latter are microscopic plants—averaging about 0.03 mm in diameter—found mainly within 60 m of the surface. They are of great importance, providing the basic food for almost all marine animal life. The pattern of primary production is affected by the availability of light and the availability of plant nutrients. Photosynthesis depends on sunlight; the depth light can penetrate sets the limits of the area in which plants can grow if basic nutrients are available. This is known as the *euphotic zone*. The main nutrient salts required for plant growth are phosphates and nitrates. Some of these nutrients are carried by rivers from the land to the ocean; this process is particularly important on the continental shelves. The bulk of nutrients, however, are provided by the decomposition of organic material in the ocean. Unlike the decomposition in soil,

Fig. 2-14. Marine ecosystem.

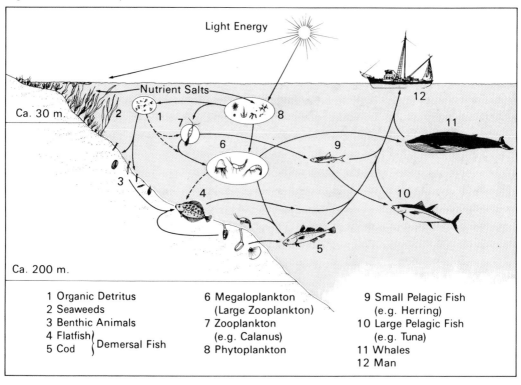

1 Organic Detritus
2 Seaweeds
3 Benthic Animals
4 Flatfish⎫ Demersal Fish
5 Cod ⎭
6 Megaloplankton
  (Large Zooplankton)
7 Zooplankton
  (e.g. Calanus)
8 Phytoplankton
9 Small Pelagic Fish
  (e.g. Herring)
10 Large Pelagic Fish
   (e.g. Tuna)
11 Whales
12 Man

which occurs within the reach of plant roots, this process in oceans generally occurs below the euphotic zone as dead organic material falls to the sea bottom. This material is returned from the sea bottom to the euphotic zone as the result of general turbulence— particularly the upwelling of cold water from considerable depths along the west coasts of continents. Without these mechanisms the upper levels of the ocean would be nearly devoid of plant life.

The production of plant material in the oceans is enormous. Because this plant material is spread over such a large area it is virtually impossible to harvest directly. People can harvest the material only after it has been concentrated by larger organisms. The process by which this occurs is referred to as a *food chain* (see Fig. 2-14). For example, phytoplankton are consumed by zooplankton (animals, 6 mm to 12 mm in diameter, with their own means of locomotion) which are eaten by larger fish such as herring, which in turn become the food for larger marine animals. As the organic material moves from one trophic level to another in the food chain, there is a rapid rate of loss. Thus a herring may consume 10 kg of plankton to increase its weight by 1 kg, while a tuna requires 10 kg of herring to increase its weight by 1 kg. This loss occurs principally because most of the food consumed is used for metabolic processes and only a small proportion contributes to the growth of the animal. It is estimated that 130 billion tonnes of plants are produced in the oceans each year. Figure 2-15 indicates the rate at which organic material is lost from one trophic level to another, and the food chain is illustrated in Fig. 2-14. This rate of depletion is of great significance to current efforts to assess the productivity of the oceans as a source of food. Whenever we take fish or plants from the various trophic levels we are interfering with the total output of a natural ecosystem.

Primary production plants (e.g. phytoplankton)	130 000
Trophic level 2—herbivores (e.g. zooplankton)	13 000
Trophic level 3—1st carnivore (e.g. herring)	2 000
Trophic level 4—2nd carnivore (e.g. tunas)	300
Trophic level 5—3rd carnivore (e.g. sharks)	45

Fig. 2-15
Estimated annual production of organic material at trophic levels, in millions of tonnes.

TYPES OF FISH

Many fish and sea mammals are found in the same locations as the largest concentrations of phytoplankton and zooplankton, principally on the continental shelves and in areas of the ocean where upwellings occur. Although there is a fantastically large number of species of fish, a simple classification is outlined here with some of the better-known species listed as examples.

THE BENTHOS

These creatures live on the sea bottom, mainly in shallow water, and include many different classes, families, and species. Such molluscs as oysters, clams, cuttlefish, and squid belong to this group, as do the crustacea, which include crabs, lobsters, crayfish, prawns, and shrimps.

THE DEMERSAL

Fish in this group live normally at depths of between 60 m and 180 m. There are two main subdivisions: the flatfish, such as flounder, plaice, sole, and halibut; and the round fish, such as cod, haddock, and hake.

THE PELAGIC

Fish of this type are usually found in shoals near the surface feeding on plankton or other pelagic fish. The most important commercially in this group are the herring, and the various members of the herring family including sardines, pilchards, sprats, and anchovies. Other major pelagic fish are the mackerel, tunny, bonitoes, and sharks.

The main commercial fish species by weight of catch in order of importance are the Peruvian anchoveta (anchovies), Atlantic herring, Atlantic cod, mackerel, Alaska walleye, pollack, and the South African pilchard. The most important invertebrates include oysters, squid, shrimps, prawns, clams, and cockles.

THE MARINE MAMMALS

These creatures are warm blooded. Some, such as whales, spend their whole life in the sea and others, such as seals, come to land to breed. Other mammals that spend most of their life in the sea include dolphins, porpoises, sea cows, sea otters, and walruses.

The Resources of the Sea

The ocean is presently important as a means of transportation, a moderator of climate, a reservoir of water vapour, and a source of food (only about one per cent of the total consumed by man). In 1972 the economic value of fish caught was approximately $15 billion (U.S.); the ocean freight bill was almost twice as great. While the value of oil and gas from the ocean floor and other minerals taken from the ocean is not known, some rough estimates for the same year indicate an economic value of at least $15 billion (U.S.). With rapidly rising energy prices in the late 1970s this situation has changed considerably.

It seems certain that the value of products from the sea—particularly economic minerals (defined on p. 264)—will grow rapidly in the future. Since only a limited amount of the ocean's potential contribution to human well-being has been realized, many scientists consider the ocean to be the last major frontier on earth. To break through this frontier sciences such as oceanography and marine biology have expanded rapidly in the past few years and will undoubtedly continue to do so in the future. Their main objectives are to develop methods of exploring and exploiting the resources of the oceans without damaging the oceans' ecosystem. The development of ocean resources will increase our knowledge of the earth and provide essential food and other materials needed by the world's rapidly growing population.

MINERALS FROM THE SEA

People have been extracting salts from the sea for many centuries. In most cases this is still accomplished by trapping sea water in shallow basins. Different compounds are then precipitated out of it through the process of natural evaporation. Such salts as iron sulphide, calcium carbonate, calcium sulphate, and the most important and abundant, sodium chloride (table salt), are recovered in this way. The remaining liquid, known as bittern, was once discarded but has now been found to contain such further by-products as magnesium chloride, potassium chloride, magnesium sulphate, and bromine, all of which can be recovered through evaporation.

It is also possible to remove certain elements directly from sea water. Areas of the ocean with unusually high concentrations of such trace elements as gold, copper, and lead are known to exist. For example, an area of water at the bottom of the Red Sea was found to contain 1000 to 50 000 times as much of the elements of iron, copper, manganese, and lead as are normally found in sea water. However the recovery of these trace elements from sea water, while possible in many cases, will depend on a future scarcity of the elements on land or on the development of economically feasible methods of processing them from sea water.

The potential for mineral exploitation is even greater on the continental shelves because the rocks of the shelves are basically similar to those of the continents. Offshore placer deposits of tin, gold, platinum, and diamonds have been discovered; sand and gravel are available in large quantities, as is glauconite, a source of potassium for fertilizer. Many other mineral deposits are known to exist although their exploitation will depend on the scarcity of these minerals on land as well as the development of new mining technologies.

On the bed of the deep-sea floor are deposits known as *manganese nodules*. Composed primarily of nickel, cobalt, lead, zinc, manganese, and copper, these potato-shaped rocks, averaging 4 cm in diameter, are not evenly distributed but occur only on specific parts of the sea bed. The richest deposits of nodules are at depths greater than 4000 m and in equatorial waters. What is particularly important about these nodules is that certain elements, notably copper, cobalt, and nickel, are accumulating on the ocean floor in the Pacific alone at an estimated rate of 6 to 10 million tonnes a year. Although the technology of extraction has been worked out, as of 1978 no mining of these nodules has been undertaken. As noted later in the chapter, no international agreement has yet been reached on the procedures to be followed by undersea mining operations.

Another type of mineral deposit, known as *metalliferous sediment*, was discovered in the Red Sea in 1948. Three closed basins with unusually warm layers of water have sea-bed sediments of yellow, red, and blue clay. These clays contain a variety of metals including copper and zinc as well as traces of silver and gold. For example, in the largest of these basins, called Atlantis II, it was estimated that the value of copper and zinc would be between $4 billion and $5 billion (1974 mineral prices in $U.S.) if the minerals could be extracted. Such deposits are believed to have been formed by undersea volcanic discharges associated with the undersea ocean ridges. Since these ridges extend throughout all the oceans, there are good possibilities that other similar metalliferous sediments will be found. But clearly there is more to undersea mineral exploration than finding the minerals. They must be mined and processed and, as noted in connection with manganese nodules, many problems must be overcome before this can happen.

Other kinds of mineral deposits lie on the sea bed awaiting utilization. These include phosphorite deposits (phosphorus is valuable as a fertilizer and is used in many branches of the chemical industry), gold, and diamonds. Oil and natural gas, of course, have been brought up from under the sea for many years in such areas as the Gulf of Mexico and the Persian Gulf. Further discoveries of oil and gas have been made under the continental shelves, and more are expected as new methods of prospecting and drilling in areas of deep water are developed.

Ocean Pollution

Many of the substances we discard either deliberately or accidentally find their way into the ocean. Even much that we discharge into the atmosphere eventually ends up in the sea. The list of these substances is very long and includes such diverse products as petroleum, concentrated sewage, chemical-warfare gases, detergents, pesticides (e.g. DDT), chemical effluents, heavy metals, and radioactive wastes. While some of the problems that these substances create are obvious, we are becoming increasingly aware that the effects of many are only incompletely understood. As the dumping of pollutants in the oceans increases, the chances that some irreparable damage may be done to the ocean environment become greater.

One of the most serious of these pollutants

is petroleum products. Estimates of the total amount released in the ocean range annually between 5 million and 10 million tonnes. The effect of major spills on ocean life has been well documented, but what is not completely understood is the long-term effect of increasing spillage. For example, we do not yet know the long-term results of the breakup of the *Amoco Cadiz* off the Brittany coast in the spring of 1978. The short-term effects of the spilling of thousands of tonnes of oil along this coast were catastrophic. (In the late 1970s there were 4000 ships transporting 11 billion tonnes of petroleum each year. More ships will certainly go down in the future.) What effect will such discharges have on the ocean's ecosystems?

Of even greater consequence is the fact that we are tampering with the earth's oxygen supply when pollutants are dumped in the ocean. Approximately 70 per cent of the world's oxygen is produced by phytoplankton through the process of photosynthesis. It has been demonstrated that in some areas herbicides and pesticides such as DDT are destroying phytoplankton. This comes at a time when oxygen-producing vegetation on land is being destroyed and oxygen consumption, particularly by the burning of fossil fuels, is rising rapidly.

In November 1972, eighty nations signed an international convention to control the dumping of waste materials into the oceans. This convention prohibits the dumping of such substances as radioactive materials, biological and chemical warfare products, various types of oils, cadmium, mercury, and organohalogen compounds (e.g. PCBS). Certain other materials such as arsenic, lead, fluorides, and pesticides can be dumped in certain quantities and only in some locations if prior permission is obtained. While this agreement has weaknesses it is generally regarded as an important breakthrough in the development of international environmental law.

Jurisdiction over the Ocean

The maritime nations of Europe agreed in 1609 to a doctrine known as 'the freedom of the high seas'. This meant that any nation could exercise sovereignty only over a narrow zone of coastal waters known as the territorial sea (see below). Beyond this zone the ships of the world could travel as they wished. While piracy and the slave trade were forbidden, fishing was free to all. In later years this freedom was extended to include the laying of cables and pipelines.

One of the authors of the doctrine of the high seas, a Dutch jurist Hugo Grotius, wrote in 1609: 'Most things become exhausted with promiscuous use. This is not the case with the sea. It can be exhausted neither by fishing nor by navigation, that is to say, in the two ways in which it can be used.' This statement was true for its time and indeed still held well into this century. The doctrine served its purpose by promoting commerce and helping to preserve peace. Today, however, it is no longer adequate. We have seen how Grotius's premise has been altered by new developments such as the use of oil supertankers, treating the sea as a dumping ground for waste materials, the mining of the under-sea floor, and overfishing. The nations of the world have recognized this dilemma. Under the sponsorship of the United Nations a series of Law of the Sea Conferences have been held in the 1970s. The principal object of these conferences has been to produce new laws through international agreement that will prevent abuses and provide for responsible management of the oceans.

THE SITUATION IN THE MID-1970S

Present international agreements recognize three different ocean zones:
a A nation has complete sovereignty over *internal waters*, including harbours, bays, and estuaries.

b A country also has complete sovereignty over its *territorial sea*, one of the zones that is presently in dispute. In the past most nations accepted a limit of three nautical miles (about 6 km). Recently, however, approximately 50 nations have unilaterally claimed twelve nautical miles, which has also been unofficially accepted. Other countries, notably Canada, the United States, and ten Latin American nations, have claimed a limit of 200 nautical miles, which they are presently enforcing. Within these territorial seas (whether legal or not) each coastal state has full sovereignty but must permit 'innocent passage' to foreign vessels. (Innocent passage is anything not prejudicial to the peace, good order, and security of the coastal state.)

c The rest of the oceans are referred to as the *high seas*. These waters are free for all to use (*res communis*) and cannot be subjected to the sovereignty of one nation. Individual nations, however, exercise jurisdiction in this area over their own ships and citizens.

Beneath the ocean surface, the ocean floor can be divided into two zones—the continental shelf and the deep-sea bottom. It is accepted in international law that coastal nations can claim exclusive right to the sea bed and the resources to be found on or below it on the continental shelves off their own coasts. Thus a nation possesses sovereignty over the sea bed on its continental shelf and no other nation can claim the resource of this area. International laws do not cover the deep-sea bottom, the area beyond the continental shelves. Traditionally it has been regarded as free for the use of all (*res communis*). However it has also been suggested that the sea bottom belongs to no one (*res nullius*). In this case it is assumed that a nation involved in using a portion of the sea bottom acquires the rights over that area during the period of use, provided there is no interference with navigation or fishing.

LAW OF THE SEA CONFERENCES

The Law of the Sea Conferences began in Geneva in 1958 and have reconvened annually since 1973. While the 148 participating nations have not yet solved the main problems, the conferences have at least defined all the issues. Some of these are noted below.[1]

1 Owing to modern fishing techniques and the lack of control over fishing, a number of serious problems have developed. Overfishing has depleted the stocks of many important commercial fish species. Because the management of fish stocks is almost impossible many species of commercial fish are virtually extinct, while many other species are endangered. The significance of this is obvious in terms of ocean ecosystems, world food shortages, and particularly the scarcity of relatively inexpensive sources of protein. In addition, overfishing by long-distance fishing nations such as the USSR, Poland, and Spain has deprived many coastal states of a resource on which much of their population depends for a livelihood. To remedy this prob-

Fig. 2-16. The approximate position of the 200-nautical mile economic zone of eastern Northern America.

lem many coastal states have unilaterally established an economic zone of 200 nautical miles (see Fig. 2-16). Within these zones they have declared exclusive sovereign rights to the management and harvesting of fish as well as regulations with respect to pollution. Many long-distance fishing nations oppose this extended coastal jurisdiction, but nonetheless it is in effect and it is being enforced. Whether it will be effective in preserving fish stocks and reducing pollution only the future will tell.

2 It is only a matter of time before commercial mining of manganese nodules on the deep-ocean floor begins. In 1970 the United Nations General Assembly declared the sea bed and its resources beyond the jurisdiction of any country. It was to be 'the common heritage of all mankind'. The operation of this concept has been debated by the annual Law of the Sea Conferences since 1973, and no international agreement has been reached on how this principle can be enforced. The main problem is related to the fact that sea-bed development will probably be undertaken first by the industrial nations. The developing countries, in particular, are concerned that they will be deprived of their share of the benefits from these resources. In addition, countries such as Chile and Canada are worried that the mining of large quantities of copper and nickel from the manganese nodules will have an adverse effect on their own production of these minerals.

3 The present regulations applying to ocean pollution are still inadequate, particularly as they apply to ships and mineral exploitation. Some coastal states feel that they should be empowered to set up and enforce their own anti-pollution standards in addition to the international standards enforced by the shipping nations. This enforcement would occur in territorial waters and for some distance beyond (for example, within the economic zone of 200 nautical miles). Many shipping nations are opposed to this latter type of regulation, fearing it might impede navigation.

STUDY 2-7

1 Examine Fig. 2-16 and explain why countries such as Canada feel that even the economic zone of 200 nautical miles may not be sufficient for adequate protection of fish stocks in certain areas. Explain the difference between the economic zone and territorial waters.

2 Many nations are seeking the adoption of an official 12 nautical mile limit to the territorial sea. The major maritime countries oppose this unless the present doctrine of 'innocent passage' is replaced by a right of 'free or unimpeded transit'. What is the difference? Using an atlas, look up the following straits and comment on the problems a limit of 12 nautical miles could create: Gibraltar, Hormuz, Malacca, and Bab El Mandeb.

3 Material related to the Law of the Sea Conferences can be updated on a yearly basis. Keep a record of published material over a period of months and comment on the problems and proposals that emerge.

4 Agreement on international regulations dealing with the ocean has been difficult to achieve because of the different attitudes among states with broad or narrow continental shelves, landlocked states, technologically advanced countries, developing countries, and countries with large merchant fleets. By having people represent countries in each of these groups a simulated Law of the Sea Conference can be set up. Research is essential to establish a particular country's position.

[1]Based in part on *The Future of the Oceans*, prepared by The Information and Legal Operations Divisions, Department of External Affairs, Ottawa, Canada.

3/ CLIMATE ELEMENTS

The atmosphere, or air, is the gaseous part of the biosphere. It provides some of the essential conditions for life. These include favourable temperature, moisture, pressure, and the oxygen needed by air-breathing creatures.

The terms weather and climate refer to the conditions of the atmosphere. These conditions include heat, moisture, and air movement. *Weather* means the day-to-day conditions of the atmosphere over a particular area and the short-term changes in these conditions. *Climate* is the aggregate of atmospheric conditions for any area over a period of years.

The conditions of the atmosphere vary greatly from place to place; at any specific place they vary during the course of a year. These variations have a strong influence on many other phenomena in the biosphere such as vegetation, soils, ocean currents, landforms, and all living creatures, including humans. Although we talk about 'atmospheric conditions' when referring to weather and climate, it is important to realize that there is a continual flow of energy and matter between the three spheres—the atmosphere, the lithosphere (land), and the hydrosphere (water). It is this interaction and interdependence that largely determine the conditions of our physical environment. Con-

sequently, an understanding of weather and climate is essential to the geographer.

The geographer's interest in weather and climate is different from that of the meteorologist and the climatologist. While they are concerned with the mechanics of atmospheric conditions, the geographer is interested in weather and climates as they contribute to an understanding of the environmental qualities of the different parts of the earth. This will help us explain the present characteristics of human occupance in different areas and also make us more aware of both the opportunities and limitations for the future development of these areas. In order to carry on such studies, the geographer must have a knowledge of some of the more fundamental aspects of meteorology and climatology. These subjects form the major part of this and the following chapter.

THE COMPOSITION OF THE ATMOSPHERE

The atmosphere is a highly compressible mixture of gases that form concentric layers around the earth. These layers vary in thickness and density (see Fig. 3-1). Traces of atmospheric gases extend up to 10 000 km above the earth's surface; 97 per cent of the total mass of the atmosphere lies within approximately 30 km of the surface. By volume, two principal gases make up the pure, dry air of the atmosphere. The first, nitrogen (78.08 per cent) by volume, is an inactive gas; a very small amount of it is extracted from the air by bacteria to form nitrogen compounds vital to plant life. The second, oxygen (20.94 per cent), is highly active chemically —for example, it combines with fuels in the process of combustion and, of course, with food to produce heat and energy in animals. Since the volume of atmospheric nitrogen is about four times that of oxygen, it serves to dilute the chemical effect of the latter. Pure, dry air also contains a number of minor gases, including argon (0.93 per cent) and much smaller amounts of neon, helium, ozone, hydrogen, krypton, xenon, and methane. Carbon dioxide (0.03 per cent) is another minor but extremely important gas. In addition to its ability to absorb solar or earth radiation— and therefore help to regulate temperatures close to the earth—it is also essential to plant life and the process of photosynthesis. Unlike gases that are constant in volume, the CO_2 content of the atmosphere is thought to be rising, mainly as a result of the enormous increase in the combustion of fossil fuels in the last few decades. As discussed later (p. 113), this increase is considered a possible factor causing changes in the world's climate.

The atmosphere is seldom, if ever, pure and dry. Its lower portions contain significant but variable quantities of *aerosols* (particles of dust, smoke, and sea-salt) and—of greater importance—water vapour. Near the surface of the earth the amount of water vapour in the air varies by volume from almost none to as much as 3 per cent. Since the quantity of water vapour in the air depends largely on the temperature of the air, the maximum water vapour values are found in the tropics. It would be difficult to overestimate the importance of water vapour to weather and cli-

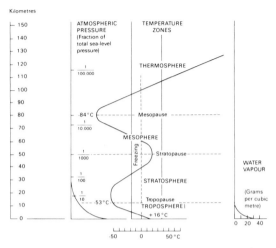

Fig. 3-1. The divisions of the atmosphere according to temperature zones.

mate. Not only is it the source of all moisture in the atmosphere, but its presence also serves to moderate the temperatures of the lower atmosphere (troposphere) by absorbing incoming solar radiation and outgoing earth radiation. In addition it is an important means by which heat is transferred from the earth to the atmosphere. (All these functions are examined in later sections of this chapter.)

The various divisions and some of the significant characteristics of the atmosphere are shown in Fig. 3-1. It is in the *troposphere*, the layer of air nearest the earth, that our weather occurs. The fundamentals of weather and climate described in this chapter and the next are associated for the most part with this section of the atmosphere. The lower layers of the atmosphere as shown in Fig. 3-1 are subdivided on the basis of temperature change. The troposphere experiences a fairly regular rate of temperature decrease known as the *lapse rate*. The average rate at which this decrease occurs, based on many observations of air temperature, is approximately 6.4°C/km. However, the actual lapse rate may vary considerably from the average at any given time over any place. In some instances the temperature may even increase with altitude for a short distance and warmer air will actually overlie colder air. This situation is called a *temperature inversion*.

The top of the troposphere, called the *tropopause*, marks an approximate altitude from 9 km at the poles to 16 km at the equator where temperatures stop declining. Within the *stratosphere* above they hold fairly constant at $-55°C$ or increase slightly. In the upper part of the stratosphere and extending into the *mesosphere* is an extremely important ozone (O_3) layer. This layer acts as a shield, absorbing ultraviolet radiation that would destroy bacteria and severely burn plant and animal tissue if it reached the earth's surface. The absorption of these rays causes the rapid increase in temperature that occurs in the lower mesosphere. Above this layer temperatures again decline until the *thermosphere* is reached. Beyond this point temperatures in the atmosphere may rise to as high as 1100°C, but temperature figures here have little meaning because the air becomes so thin that little heat can be either held or conducted.

In order to understand the patterns of climate over the world it is first necessary to understand the nature and cause of variations in heat, moisture, and air movement. These three conditions of the atmosphere are discussed next in this chapter. Following this, the effect of these factors on the global distribution of temperature and precipitation is examined. Then, in Chapter 4, a classification of climate is used to examine in more detail the patterns of climate over the earth.

STUDY 3-1

1 Why is climate defined as the aggregate rather than the average of atmospheric conditions?
2 Explain what is meant when the atmosphere is described as a highly compressible mixture of gases.

3 Examine the following temperature distributions. Which is the closest to the average lapse rate? Which represents an inversion?

	ground	300 m	600 m	900 m	1200 m
a	18.0°C	20.0°C	16.5°C	14.0°C	12.5°C
b	18.0°C	16.0°C	14.5°C	12.5°C	10.0°C
c	18.0°C	15.5°C	12.5°C	10.0°C	9.0°C

4 Discuss some of the ways in which people have overcome such limitations of weather and climate as inadequate precipitation or temperatures that are too low or too high.

SOLAR ENERGY

If the sun suddenly burned out, the temperature at the earth's surface would fall to almost absolute zero ($-273°C$). Life would cease to exist. Although some heat flows through the earth's crust from its interior, the amount is insignificant. Thus we can state that all life processes on our planet are dependent on energy radiated by the sun.

The earth receives about one two-billionth of the sun's total output of energy. The flow of energy from the sun and back again into space involves a rather complex series of processes. In the first place it is important to realize that the energy received by the earth is equalled by the loss of energy from the earth to space. This concept of a *radiation balance* is rather obvious since in the short run (months and years) the outside of the earth and the atmosphere are not getting warmer or colder. (In the long run—decades and centuries—changes do occur, and these are discussed on pp. 109 to 115.) However, the amount of solar energy that is gained or lost varies over the earth's surface and in any specific place changes from season to season and from day to night. For example, the lower latitudes receive more energy than is lost while the opposite is true for the polar latitudes. (Fig. 3-2). To correct this imbalance there are mechanisms that transfer heat from the lower to the higher latitudes. These are

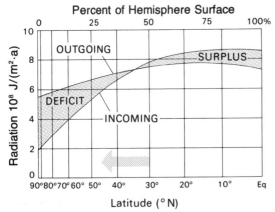

Fig. 3-2. An illustration of the balance between in-coming solar radiation and outgoing earth radiation. What do we know about the relationship between the areas marked 'surplus' and 'deficit'? The zones of pernament surplus and deficit are maintained in equilibrium by the transfer of energy poleward. How is this transfer carried out?

the moving air and water currents of the atmosphere and oceans.

Not all the energy received by the earth is involved in heating the atmosphere before being lost to space. As noted in Chapter 1, a small part (about 5 per cent) is absorbed by green plants through the process of photosynthesis to produce organic molecules that are used in the growth of the plant. Much of this energy is eventually lost to the atmosphere through the process of respiration, and less than 1 per cent of the energy reaching the earth actually ends up as plant material. The plant material, however, includes all the food used in the biosphere by plants, animals, and humans.

Some of the solar energy is stored in other ways, for example, in the soils and trees of the forest. In fact, even the energy we derive from petroleum, natural gas, and coal is the solar energy used by plants and animals millions of years ago.

We all know that the amount of solar energy—as reflected in the temperature—varies from place to place. The importance of these differences cannot be exaggerated. Everyone can think of commonplace examples related to temperature differences. Some of the more obvious include the different plant and animal communities, various types of agriculture, special characteristics of buildings, recreation, clothing, and even the shape of landforms. We are so accustomed to these temperature differences that most people could provide a fairly accurate assessment of the temperature characteristics of a place from a photograph showing some of its natural or man-made phenomena. It is obviously important, therefore, that we understand how these differences are caused.

Heating the Atmosphere

Solar energy is properly described as *electromagnetic radiation* but commonly known as *solar radiation* or *insolation* (a contraction of incoming solar radiation). It is made up of rays ranging from the invisible infra-red and ultra-violet rays at either end of the solar spectrum to rays from the visible portion of the spectrum between these extremes, which we see as sunlight. Travelling at the speed of light (300 000 km/s), it takes over eight minutes to reach the earth. A celestial body such as our sun emits a quantity of radiation proportional to its temperature. The wavelength of this radiation is inversely proportional to its temperature, that is, the higher the temperature the shorter the wavelength. It follows that the hottest bodies not only radiate the most energy but most of this radiation occurs in the ultra-violet or shortwave portion of the electromagnetic spectrum. So-called cool objects like our own planet—indeed also our own bodies—radiate in accordance with their temperature, but this is mainly in the infra-red or long-wave part of the spectrum. The significance of this is made clearer in the following paragraphs.

Upon reaching the atmosphere the incoming solar radiation is received in the following

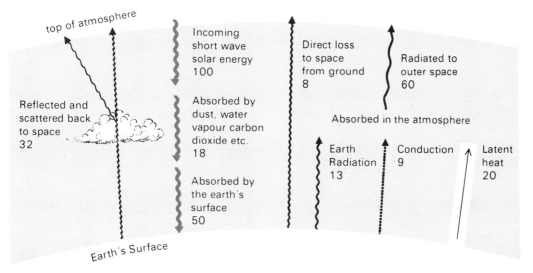

Fig. 3-3. The incoming solar radiation at the top of the atmosphere is represented as 100 units. The energy transfers are explained in the text with the exception of the direct loss of 8 units from the earth. This loss occurs through 'radiation windows', bands in the spectrum where absorption does not occur.

ways (see Fig. 3-3). On the average 17 per cent of all solar radiation is absorbed by the water vapour, ozone, clouds, and dust in the atmosphere. Approximately 32 per cent is reflected back by clouds and other reflecting surfaces, including the earth, or is scattered out into space and has no effect on the atmosphere. The actual amount reflected depends on the quantity and type of cloud and characteristics of the earth's surface. The amount of solar energy reflected by the surface is called the *albedo*. Snow and ice reflect back a high degree of radiation, while water absorbs a large amount of the radiation it receives; forest and farmland generally vary between these extremes. The remainder of the solar radiation that is not reflected or scattered heats the earth's surface. This heat, in turn, is continually radiated back into the atmosphere. However, since the earth's average surface temperature is much lower than that of the sun, its radiation is consequently of a longer wavelength. The earth's

long-wave radiation is absorbed more readily by the atmosphere than incoming short-wave solar radiation is absorbed by the atmosphere. As much as 60-70 per cent of the long-wave earth radiation is retained while the rest escapes into space. Water vapour and carbon dioxide are the principal absorbing gases. Thus, as a radiating body the earth is the main source of heat for the lower part of the atmosphere.

While some of the energy absorbed by the lower atmosphere is radiated into space and lost, more of it is radiated back towards earth. The retention of this energy in the atmosphere is extremely important. If it did not occur the average temperature on the earth's surface would drop considerably both at night and during the winter season. For example, the heat loss from the earth's surface on a clear night is much greater than the loss on a cloudy night. This is ample evidence of the particular importance of water vapour in preventing heat escape. As we noted at the beginning of this section, however, over a period of time the loss of heat from the atmosphere balances the total amount of incoming radiation and the earth maintains a radiation balance.

In addition to radiation, the atmosphere is heated by the release of heat through the process of condensation and conduction. The more important of the two is the heat released to the atmosphere as condensation occurs. This happens when the solar energy received by the earth is used to evaporate water. The energy used in evaporating the water vapour is held in the water vapour in latent or potential form. It is known as *latent heat*. As the air containing this water vapour is forced upwards, it cools, the water vapour condenses, and the latent heat is released to the atmosphere.

Conduction of heat occurs when two bodies with different temperatures come in contact. Heat flows from the warmer to the cooler body until the same temperature is attained by both or until the contact is broken. For example, when land is being heated during the day some of this heat is transferred to the lower layers of the air by conduction. This process is aided by convectional currents and general air turbulence, which help spread through the troposphere the heat acquired by surface air.

Although the solar energy received by the atmosphere is balanced by the energy it loses, there is a marked difference in the rate of gain and loss at different latitudes and at the same latitudes at different seasons. Generally the amount of incoming solar radiation exceeds the amount of outgoing earth radiation in the lower latitudes, while the reverse occurs in the higher latitudes (Fig. 3-2). To correct this imbalance the motions of the atmosphere and the oceans act as mechanisms of heat transfer, exporting heat from the equatorial to the polar regions. In fact, this energy imbalance is the prime cause of the earth's atmospheric and oceanic circula-

tion. Approximately 80 per cent of the energy transfer is accounted for by winds in the atmosphere; the remainder by ocean currents. In turn, these transfers are among the most important factors giving rise to variations in weather and climate over the earth. They are discussed further in the section on atmospheric pressure and winds.

Latitudinal Variations in Insolation

The general decrease in temperatures from the equator to the poles is one of the most commonly known facts of climatology. This decrease results from variations in the amount of insolation received at the earth's surface. As described on pages 6 to 10, there are three main factors responsible for the amount of insolation received at any place:
1 the duration of the period of daily sunlight;
2 the intensity of the sun's rays, i.e. the angle at which they strike the earth;
3 the transparency of the atmosphere.
The transparency of the atmosphere varies according to the amount of water vapour, dust, and clouds it contains. Areas that experience a large number of cloudy days, or cities with polluted air, will receive less direct insolation since a large proportion of the incoming radiation will be reflected or scattered back into space.

Different Heating Characteristics of Land and Water

Although the amount of insolation received is roughly the same along any parallel of latitude, there is often a considerable difference between the temperature of land and water at the same latitude. This results in a difference in the amount of heat transferred to the atmosphere from each of these surfaces. The reason is that land heats and cools much more rapidly than water, whose currents keep it constantly in motion. Consequently solar energy must heat a great volume of water but only little more than the surface of the land. Other reasons for the different heating characteristics of land and water include the facts that (a) solar radiation penetrates deeper into the water and (b) the specific heat of water is higher than that of land (a quantity of water heats up much more slowly than a similar quantity of dry land).

In summer, then, temperatures will be considerably higher over a large land area than over a large body of water at the same latitude. This temperature difference would be of little importance if the atmosphere were motionless, but since air moves, the temperature characteristics of one area are often transferred to another. In many cases the land areas adjacent to large bodies of water are influenced by air that has moved from water to land, bringing with it temperature characteristics picked up over the water. A more detailed explanation of these air movements and their importance will be discussed in the section on 'Air Masses' later in this chapter.

STUDY 3-2

1 Factors (1) and (2) in the previous section on 'Latitudinal Variations in Insolation' have already been discussed in the sections on 'Day and Night' and 'The Seasons' in Chapter 1. Review this section and summarize the influence of the intensity and duration of solar radiation by explaining the information shown on Fig. 3-4.

2 What would the temperature characteristics of the earth be like if there were no atmosphere? Explain why the daily variation in temperature is much greater in desert areas than in more humid areas at the same latitude.

3 a Examine and explain Fig. 3-3, which illustrates what happens to the solar radiation that enters the atmosphere.

b Why is the influence of the atmosphere

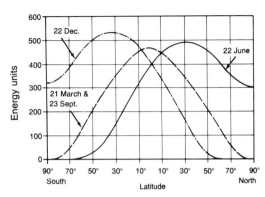

Fig. 3-4. The latitudinal distribution of solar energy over the earth's surface at the equinoxes and solstices.

on solar and earth radiation called the 'greenhouse effect'?

c Give some other practical examples of the greenhouse effect from your own experience.

4 Referring back to Fig. 3-1, note how far into the atmosphere temperatures decrease with altitude. Why do temperatures in the upper atmosphere *increase* as you move away from the earth?

5 While temperatures normally decline as altitude increases in the troposphere, it is not uncommon to find just the opposite, that is, a temperature inversion. Because the land surface is a more efficient radiator of heat than the air, one of the most common types of inversions occurs near the ground, usually at night. An example of such a radiation inversion is shown in Fig. 3-5.

a Conditions favourable to temperature inversions near the ground include long winter nights, clear skies, dry air, calm air, or a snow-covered surface. Explain how each of these or combinations might lead to a temperature inversion.

b Explain why temperature inversions can result in serious hazards, particularly when they occur over large cities. Why is there

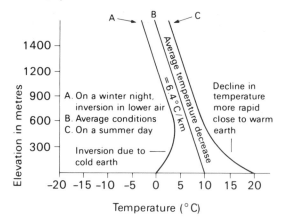

Fig. 3-5. The vertical distribution of temperature under different conditions (after Trewartha, Robinson, and Hammond).

often a close relationship between surface temperature inversions and fog and frost?

6 What are some of the temperature characteristics of (a) land-controlled climates, and (b) water-controlled climates? Referring to Figs. 3-25 and 3-26, explain why the west coasts of continents have a more water-moderated climate than east coasts.

7 On about 22 December the south pole receives its maximum radiation for a single 24 h period. When does the north pole receive its maximum radiation? Why are they not quite identical quantities? When does the equator receive its maximum radiation? Why is this less than the maximum quantity received at either pole?

ATMOSPHERIC MOISTURE

Of all the gases that make up the atmosphere, water vapour is the most variable in volume. Although only a minute fraction of the earth's water (Fig. 2-1) is stored in the atmosphere at any one time, water vapour is the most important gas as far as weather and climate are concerned.

In addition to the direct relationship between precipitation and water vapour, this gas is also important for several other reasons.

1 As we have seen, it absorbs both incoming solar radiation and energy radiated from the earth, and in this way helps regulate temperatures.

2 Through the process of evaporation it conveys latent heat into the atmosphere. The greater the amount of water vapour, the greater the amount of latent energy the atmosphere will be capable of retaining. The retention of latent energy may lead to the development and growth of atmospheric disturbances or storms.

3 Water vapour is the prime factor affecting humidity, cloudiness, and visibility.

4 Together with heat, fresh water is an essential ingredient for plant and animal life. Through the process of evaporation water vapour is transferred to the land and falls as precipitation.

Measuring Water Vapour

Water vapour in the atmosphere is called *humidity*. The amount of water vapour that the air can hold depends almost entirely on the temperature of the air, as Fig. 3-6 indicates. There are two common ways of stating the quantity of water vapour in the air. *Absolute humidity* refers to the actual amount of water vapour in a specific volume of the atmosphere. This means the weight of water vapour per cubic unit of air and is stated in grams per cubic metre (g/m^3). *Relative humidity* refers to the proportion of water vapour present in the air compared to the maximum amount of water vapour possible at the same temperature. It is usually expressed as a percentage. Thus, by definition, *saturated* air has a relative humidity of 100 per cent. The temperature at which air becomes saturated as a result of cooling and at which condensation normally begins is known as the *dew point*. (If the temperature of the air is below freezing, it would more properly be called the *frost point*.)

Fig. 3-6.
Maximum water vapour holding capacity of one cubic metre of air at different temperatures (approximate).

Temperature °C	Water Vapour grams per cubic metre
− 15	1.4
− 10	1.9
− 5	3.0
0	4.9
5	6.9
10	9.4
15	12.6
20	17.3
25	23.0
30	30.4
35	39.8

STUDY 3-4

1 Using Fig. 3-6, determine how many more grams of water vapour a cubic metre of air can hold at 5°C than at 0°C? How many grams more water vapour can a cubic metre of air hold at 35°C than at 30°C? What general conclusion can be drawn about the rate of increase of air's capacity to hold water vapour in relation to increasing temperature?

2 What is the relative humidity of a cubic metre of air that contains 3.0 g of water vapour at 20°C? How much would the temperature have to fall before the relative humidity reached 100 per cent? What happens at this point?

3 If the relative humidity of the air was 73 per cent and the temperature 25°C, find the absolute humidity, the dew point, and the amount of water vapour that would be condensed if the temperature fell to 15°C.

Forms of Condensation

Fog, dew, and frost are relatively common but minor forms of condensation. They are caused when shallow layers of air next to the ground, a lake, or the sea are cooled to their dew point or below through either radiation or conduction. Fogs may also result from the movement of warm, moist air over a cold surface. In this situation the air layer closest to the ground loses heat to the ground; if the temperature drops below the dew point, condensation in the form of fog will result. Such fogs are referred to as advection fogs. Fog can be a major environmental hazard, particularly in its effects on transportation.

Of much greater importance to life on earth is the condensation that takes place well above the earth's surface in large, upward-moving masses of air that cool, condense, and form clouds. It is from this process that all precipitation stems. When the temperature of air reaches the dew point (owing to the movement of air upwards from the earth) and condensation begins, each droplet of water condenses around a microscopic particle of some solid substance, called *condensation nuclei*, in the atmosphere. These may include dust particles, common salt, or sulphuric acid. The visible effect is clouds. All clouds are made up of very small droplets of water or minute crystals of ice so light that they can be sustained in the atmosphere by only a very slight upward movement of air. (Under natural conditions nuclei are never in such short supply as to hinder condensation, although in the laboratory extremely clean air can be cooled to well below its dew point without the water vapour condensing.) Some of the major types of clouds are shown on Fig. 3-7.

Of course, condensation in the atmosphere in the form of clouds does not *always* result in precipitation. In order for precipitation to occur, the small droplets of water must join together in drops too large to remain sus-

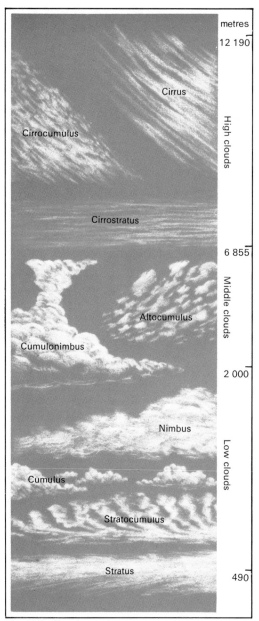

metres

12 190

Cirrus

Cirrocumulus

High clouds

Cirrostratus

6 855

Middle clouds

Altocumulus

Cumulonimbus

2 000

Nimbus

Low clouds

Cumulus

Stratocumulus

Stratus

490

Fig. 3-7. Cloud types.

pended. This probably takes place most readily when ice and water particles are present
together in the same cloud. The seeding of
clouds with dry ice or silver iodide to force
rain is based on this theory.

Precipitation

Precipitation refers to water that falls to the
earth in either a liquid or a solid state: rain,
snow, sleet, and hail.

Rain is by far the commonest and most
widespread form of precipitation. Snow, or
solid precipitation, develops in air below the
freezing point. (The equivalent of one millimetre of rain is normally ten to twelve millimetres of snow.) Sleet is frozen or partly frozen
raindrops, while hail is caused by the super-
cooling of particles of water in convectional
updrafts. Each hailstone is formed by layer
upon layer of moisture condensing around a
particle of ice; hailstones vary in size according to the intensity of the convection that
generated them. Both sleet and hail occur
infrequently, but can be extremely destructive. However, the destructive power of sleet
may not be readily recognized because sleet
is often referred to as 'freezing rain' when it
brings down power lines, branches of trees,
and so on. Sleet may also fall harmlessly
when the temperature at ground level is
above freezing.

Precipitation must be preceded by either
condensation or sublimation, or a combination of the two, and all precipitation is the
result of rising masses of air and the cooling
of this air below the dew point.

When air rises, either by force or spontane-
ously, there is less pressure on it at the new
altitude and it expands. However, to make
room for itself as it expands, it must displace
other air. The work of displacement requires
energy, which is taken out of the expanding
air in the form of heat. As a result the temperature of the rising air is lowered even though
no heat energy is lost to the air outside. Thus,
rising air expands and cools while, in the
opposite situation, descending air is compressed and becomes warmer. The rate of
cooling or heating that results from this verti-
cle movement of air is called the *adiabatic
rate* (from the Greek, 'without passage

through' meaning that cooling or heating is a result of processes internal to the air mass rather than as a result of loss or gain of heat to the surrounding air). For unsaturated air (air in which no condensation is occurring) the rate is constant at approximately 1°C/100 m. This is known as the *dry* adiabatic rate. For saturated air (air moist enough to lead to condensation, thus releasing heat) the rate of temperature change will be less, and although it varies depending on the temperature of the air, an average value is 0.6°C/100 m. This is referred to as the *saturated* adiabatic rate.

It is important to emphasize the distinction between the adiabatic rate and the lapse rate. The adiabatic rate refers to changes of temperature in rising or falling air due to expansion or contraction of such air. The lapse rate is simply a statement of the change of temperature at different levels in the lower part of the atmosphere—at a particular place at any given moment—*provided that this air is not rising or falling.*

Almost all precipitation results from the cooling that occurs with the expansion of rising air currents.

CAUSES OF RISING AIR CURRENTS

Convectional, orographic, and frontal are terms used to describe the three means by which air is forced to rise. These may be thought of as triggering devices that merely give the surface air an initial push upwards. It should be noted that much precipitation is a result of the combined effects of more than one of these triggers.

CONVECTIONAL

In tropical areas, and also over large land masses in higher latitudes, the surface air is heated and spontaneous vertical updrafts are set in motion within a convectional cell. (Convectional cells are the systems of rising and

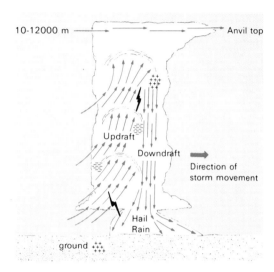

Fig. 3-8. Diagrammatic representation of the interior of a thunderstorm cell showing its anvil-shaped head and positive and negative electrical charges.

falling air by which heat is transferred within the atmosphere. This process is called convection.) This often occurs on hot summer afternoons when the surface of the earth has become unusually warm. Condensation resulting from convection usually results in cumulus or towering cumulonimbus clouds and is often sufficient to cause precipitation in the form of thunderstorms (see Figs. 3-8 and 3-9). This precipitation, although usually heavy (and in some areas torrential), seldom lasts very long and in most cases affects only a relatively small area. It is the rapidity with which air rises in a convectional updraft that results in particularly heavy precipitation and dangerous turbulence—even for the largest jetliners. Figure 3-12 illustrates the process of convection.

OROGRAPHIC

A change in elevation may be sufficient to force the upward movement of air (Fig. 3-10). (The term orographic means 'relating to

Fig. 3-9. Rain falling from heavy cumulonimbus clouds.

mountains'.) The influence of large mountains in causing air to rise and precipitation to occur is obvious from a comparison of Fig. 3-27, showing the world distribution of precipitation, and Fig. 9-1, the world map of landforms. For example, the prevailing westerly winds that sweep across the Pacific bring moist, cool air to the coast ranges of central and northern California, and heavy rainfall results on the windward slopes of these ranges. Yet the same air is drier and heated adiabatically as it descends the leeward side of these same ranges. Such dry warm winds may cause rapid temperature changes on the windward sides of mountains. They are known as *chinook* winds in North America and *foehn* winds in Europe.

It should be noted, however, that much smaller landforms may cause orographic air ascent. Over a land mass during the summer, orographic updraft may be triggered by a very small change in elevation. Often the effect of an elevated area is to intensify the precipitation associated with convectional or frontal air ascent.

FRONTAL

Where two bodies of air flowing in different directions and with different temperatures and densities meet, the warmer, lighter air will be forced to rise. The plane or boundary between the unlike air masses is known as a *front* (Fig. 3-11).

In tropical areas, where the opposing air currents have similar temperatures, the lifting is usually vertical and is accompanied by convection. In middle latitudes, frontal convergence is associated with a more gradual ascent of warm air. In this situation, particularly when the warm air is pushing the cold air back at the same time, it is being forced to rise along a very gradual slope. Condensation occurs slowly, and the resulting clouds and precipitation may be very widespread. In middle latitudes in the cooler months of the year

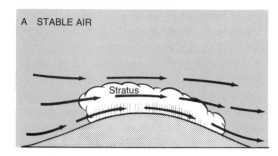

A STABLE AIR

Stratus

B UNSTABLE AIR

Cumulus

Fig. 3-10. What determines the amount of precipitation experienced on the windward and leeward side of any landform barrier? (The different effects of stable and unstable air crossing the same range of mountains are explained in the section on stability.)

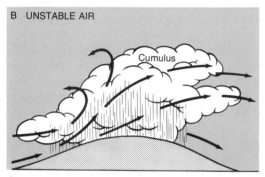

that will rise of its own accord even after the initial triggering force is no longer operating.

Figure 3-12 illustrates a stable air body. If it is forced to rise by one of the triggering devices already described, at 300 m it will have cooled to 27°C at the dry adiabatic rate of 1°C/100 m. The surrounding non-rising air will be only 1.9°C cooler than the air at the surface (assuming the average lapse rate of 6.4°C/km). The rising air is therefore cooler and heavier than the surrounding air. Unless some very strong force pushes it upwards, the rising air will tend to sink back towards the ground and little or no condensation will occur. We often experience days in which this type of condition prevails, with sunny conditions and small, puffy cumulus

a locality may experience several days with continuous cloud cover and intermittent drizzle as a result of this type of upward movement. This form of precipitation is examined in more detail in the section on 'Air Masses, Fronts, and Storms' (p. 71).

STABILITY

Stability is defined as the ability of an object to return to its original position after having been subject to some disturbance or displacement. Meteorologists use this term to describe the condition of air. When air is forced upwards, some bodies of air may resist upward movement and eventually sink back to lower levels. This air is said to be *stable*. Conversely, an *unstable* mass of air is one

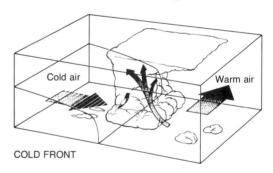

COLD FRONT

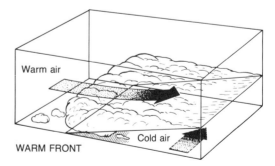

WARM FRONT

Fig. 3-11. Along a cold front the warm air rises rapidly, and, if condensation occurs, thick cumulus clouds usually result. Along a warm front the warm air rises over the cold at a very gentle slope (1 in 150). If condensation occurs, widespread horizontal formations of stratus clouds result.

clouds. On such days these clouds would dis-solve as fast as convectional updrafts prod-uce them.

Figure 3-12 illustrates an unstable air body. The sun has heated the surface of the earth, which in turn has substantially warmed the lower layers of the air. The result is a lapse rate considerably larger than the aver-age of 6.4°C/km. The body of air near the ground is forced to rise convectionally because it is lighter than the less intensely heated air over nearby ground. As it rises, this body of air is cooled adiabatically. At 300 m the temperature of the rising air is still above the temperature of the surrounding air and so it will continue to rise. At 700 m the dew point is reached and, since latent heat is being released through the process of con-densation, the air is now being cooled at the lower, saturated adiabatic rate. This unstable air will continue to rise as long as it is warmer than the surrounding air. The rising process is

usually terminated when most of the water vapour is condensed and the dry adiabatic rate then quickly reduces the temperature of the rising air below that of the surrounding air.

In general, therefore, when the lapse rate is above average and the air is humid, a con-dition of instability prevails. A small or weak lapse rate, especially one where the tempera-ture increases with altitude (an inversion), indicates stable air. Stability is important because it is linked with the amount of pre-cipitation. Abundant precipitation is much more likely from an unstable air mass than from a stable one, although naturally the water-vapour content of the air, together with the kind of triggering influence involved, must be taken into consideration. Air stability is also an important factor in air pollution, espe-cially in urban areas. Stability determines how readily air will rise, thus enabling air currents to transport smoke and other pollutants.

Fig. 3-12. Stable and unstable air conditions (after Strahler).

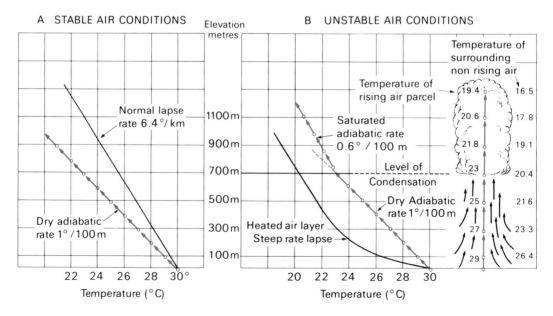

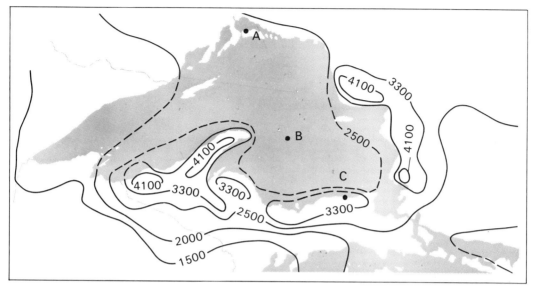

Fig. 3-13. Snowbelts along the south shore of Lake Superior. Figures indicate snowfall in millimetres.

STUDY 3-5

1 A westerly air stream from the Pacific Ocean is approaching place A (elevation 400 m), rising up the windward side of a mountain range to the top at place B (elevation 2000 m), and down the leeward side past C (elevation 700 m). The following conditions apply to this air stream.

  1 Wet adiabatic rate = 0.6°C/100 m.
  2 Dry adiabatic rate = 1°C/100 m.
  3 Initial condition of the air at sea level: stable, temperature 20°C, and dew point 10°C.

Using the information above as well as Fig. 3-6, find the following information.

  a The relative humidity at A.
  b The altitude where condensation first occurs.
  c The absolute and relative humidity at B.
  d The amount of moisture condensed by the time this air mass has reached B (in grams per cubic metre).
  e The temperature and dew point at C.
  f The absolute and relative humidity at C.

2 Fig. 3-13 shows the snowbelts on the windward side of bodies of water such as the Great Lakes. The following conditions apply to a north-westerly air stream crossing Lake Superior.

  1 The date is 30 November.
  2 The surface temperature at A is −15°C.
  3 The lapse rate of an air stream passing over A is 0.3°C/m.
  4 The water temperature is 2°C.

    a Make three graphs to show the lower 500 m of this air stream at A, B, and C. Make your own estimates of what the air temperature at the surface would be at B and C.
    b How do the passage of the air stream over the lake and the changes in air temperature explain the presence of the snowbelts, particularly on the Michigan shore? Why are the snowbelts generally quite narrow? Would you expect heavier snowfall in November or February? Explain.

ATMOSPHERIC PRESSURE AND WINDS

Cool air in northern Canada is expected to advance rapidly south-eastward today, covering the north of the province by morning and southern parts by the evening. Generally cool temperatures are expected tomorrow with variable cloudiness and scattered showers in the morning.' This daily weather synopsis from a North American newspaper could well apply to many parts of the world. It illustrates that weather and climate can be fully explained only in terms of the movements of the atmosphere. Air is constantly taking on temperature and moisture characteristics in one area and then moving elsewhere. Changes in temperature and precipitation, particularly in the middle latitudes, result from these air movements. Although the circulation of the atmosphere is not fully understood, it has been clearly established that movements of air result directly from small variations in pressure. Consequently, before examining the world pattern of winds, it is necessary to know something about atmospheric pressure and how it varies over the earth's surface.

Atmospheric Pressure

Atmospheric pressure is quite simply the weight of the atmosphere at any given point. The mixture of gases that comprise the atmosphere exert pressure on the surface of the earth. For example, at sea level a column of air extending to the top of the atmosphere and one square centimetre in cross section would exert a pressure of about 1 kg. Since this external pressure is balanced by an equal pressure of air within liquids or hollow objects (including people), pressure is not something that we commonly notice. The metric unit used to measure atmospheric pressure is the kilopascal (kPa) and is employed on the maps throughout this section. Average sea-level

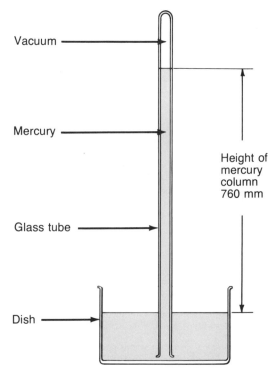

Fig. 3-14. A simple mercury barometer.

pressure is approximately 101.3 kPa, and variations from this average, although they are very important, are seldom more than 4.0 kPa either way. The mercury barometer (graduated in millimetres) is the most accurate instrument for measuring atmospheric pressure, and most other instruments are checked against it. A simple barometer, similar to the one invented by Torricelli in 1643, is shown in Fig. 3-14.

STUDY 3-6

Figure 3-14 shows a simple mercury barometer.
1 Explain how it works.
2 Explain why mercury is used instead of water.
3 Average sea-level pressure is 760 mm of mercury. What does this mean? This mea-

surement is also referred to as the *standard atmosphere* (atm). This is measured at sea level at 45°N or 45°S and is equivalent to 101.325 kPa.

4 Most barometers are of the aneroid type. Explain how they operate. Explain how such an instrument can also be used as an altimeter.

WORLD DISTRIBUTION OF PRESSURE

Figures 3-17 and 3-18 show the distribution of winds in the lower atmosphere and of pressure at sea level for the world during January and July. It must be emphasized that each of these maps shows *average* conditions for these months; considerable variation might be seen on a wind and pressure map for any

Fig. 3-15. Generalized continent.

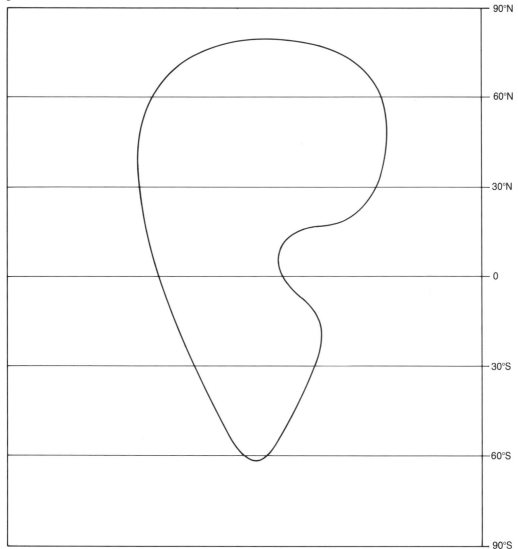

one day (Fig. 3-21). The lines on the maps are called *isobars* and join places having the same atmospheric pressure; arrows show prevailing wind direction. Figure 3-16 shows two profiles of atmospheric pressure, which represent the average pressure for all meridians from pole to pole. Figures 3-16, 3-17, and 3-18 are the basis of the following study, in which the world pattern of pressure is examined.

STUDY 3-7

1 Figure 3-15 represents a generalized continent. Draw two enlarged and identical copies of this diagram, marking on them the parallels of latitude as shown. Give one the title JANUARY and the other JULY.

2 a Using Figs. 3-16 and 3-17, print H on the copy of the January map where pressures are relatively high and L where they are relatively low. Because Fig. 3-15 resembles North and South America, it would be best to take your examples from these continents and the adjoining oceans. Locate these highs and lows as accurately as possible in relation to the parallels of latitude. Draw one or two isobars around each of the highs and lows to illustrate the pattern of isobars.

b Using Figs. 3-16 and 3-18, show the pressure systems for July in the same way.

3 The following questions are based on the two simplified pressure maps prepared as above.

a What features of pressure are common to both maps? Label the pressure areas common to both maps: the Polar High, Sub-Polar Low, Sub-Tropical High, and Equatorial Low.

b To establish the seasonal variations in pressure, compare the January and July maps, noting (i) any change in the latitudinal position of the pressure areas, (ii) changes in pressure over the land mass, and (iii) the pressure reading in the sub-tropical high over the ocean to the west of the land mass (or any pressure areas that are common to both seasons). Use this information to write a note on the seasonal variations in pressure.

4 Variations in pressure are very complex and result from many different causes. In some areas air rises and creates areas of low pressure, while in others air may be settling, resulting in high pressure. Some of the pressure areas may be thermally induced, that is, they may be a result of variations in the heat energy received at the earth's surface. What pressure areas are thermally induced? What other pressure areas are not adequately explained by variations in temperature? How are the seasonal variations in pressure related to changes in temperature?

5 The patterns formed by the isobars describe different pressure situations and are referred to as cells, troughs, or ridges. Explain these terms and find examples of each.

6 It is important to appreciate that the words 'high' and 'low', when applied to pressure, are entirely relative. Explain why this is so.

WORLD DISTRIBUTION OF WINDS

The movement of air is caused by differences in atmospheric pressure as air moves naturally from areas of high pressure to areas of low in an attempt to equalize pressure. Vertical air movements are usually described as updrafts or downdrafts. Winds can be defined as the horizontal movement of air.

The velocity of winds is controlled by the *pressure gradient*, that is, the rate and direction of pressure change. Just as some land is high and slopes to lower areas, so areas of high pressure are separated from areas of lower pressure by gradients of varying steepness. Thus isobars are similar in principle to contours. When isobars are close together the pressure gradient is steep and winds are strong.

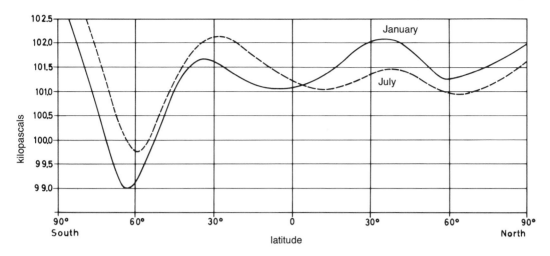

Fig. 3-16. Profiles of sea level pressure from pole to pole, showing the average for all longitudes.

A simple example of this relationship is the phenomena of land breezes and sea breezes that commonly occur along coasts. On a sunny day the land surface heats up rapidly, which also causes the lower layers of air to be warmed. The warm air expands and becomes less dense. As a result air pressure over the land will be relatively lower than that over the cooler water surface. This establishes a pressure gradient from water to land that sets up a sea breeze. At night the opposite may occur with the air over the land cooling, contracting, and becoming denser than that over the water. Since temperature differences at night are usually not as great between land and water, the pressure gradient is less and consequently the land breeze is usually weaker.

Winds are identified by the direction from which they blow. Thus a wind coming from the south west is travelling towards the north east but is described as a southwest wind. Wind velocity is measured in kilometres per hour by an instrument called an anemometer.

STUDY 3-8

This study is similar to the previous one on

pressure. The same two maps that were prepared to show the pressure systems can be used to illustrate atmospheric circulation.

1 Referring to Fig. 3-17, examine carefully the relationship of winds to the pressure areas. Start with the sub-tropical high and draw a similar but simplified pattern of winds on the January map. Use continuous lines or arrows extending from the high-pressure area into the adjoining areas of low pressure (both the sub-polar and the equatorial lows). Note that the winds do not trace a straight path but are deflected.

Do the same with the winds diverging from the polar highs. When the map is finished, the pattern of prevailing winds should be clear for every area of the map.

2 Referring to Fig. 3-18, complete the wind system on the July map in the same way.

3 a What is the direction of air movement in the low latitudes (0°-30°) and the middle latitudes (30°-60°)?

b Complete the labelling of the diagram by including the following names given to prevailing winds: Northeast and Southeast Trades, Westerlies (both hemispheres), and Polar Easterlies (both hemispheres).

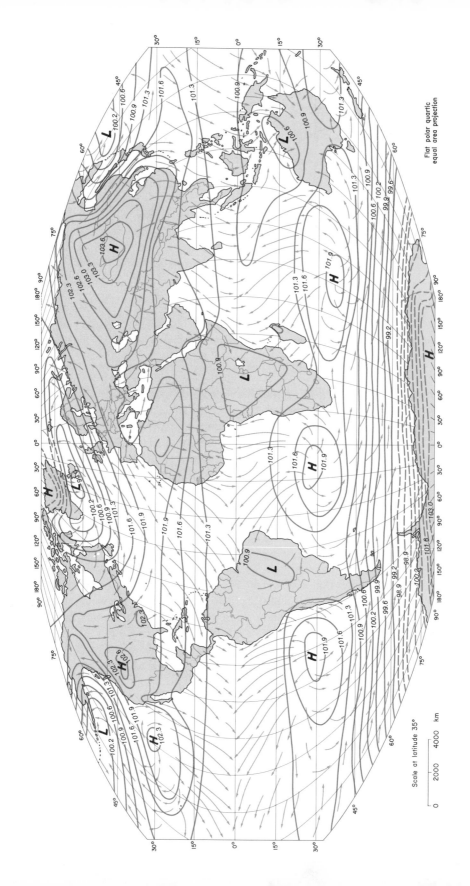

Fig. 3-17. Average sea level pressure and winds, January.

Flat polar quartic
equal area projection

Scale at latitude 35°

0   2000   4000   km

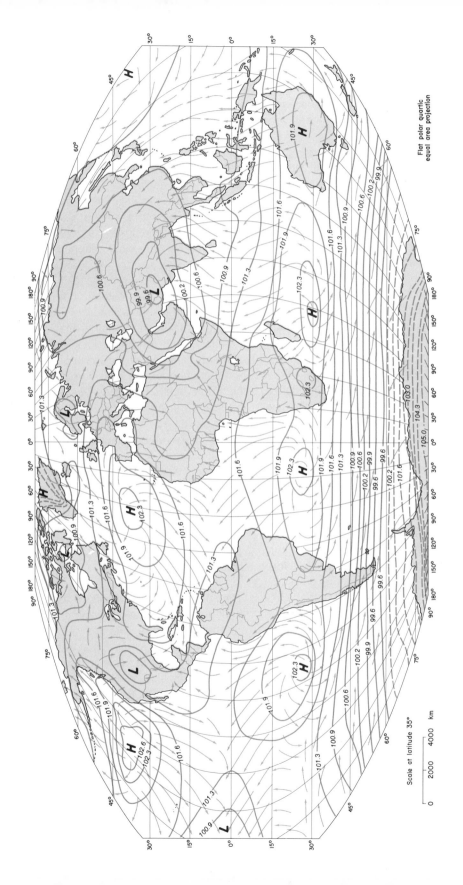

Fig. 3-18. Average sea level pressure and winds. July.

Flat polar quartic
equal area projection

Scale at latitude 35°

0   2000   4000   km

4 Can you discover any major variations in the pattern of winds between the January and July maps? Account for any differences you note.

5 Using Figs. 3-17 and 3-18 as sources, make four sketches representing four simple cellular pressure systems—a high- *and* low-pressure cell for each hemisphere. Draw in the winds and note whether each is (a) diverging or converging, or (b) rotating clockwise or counter-clockwise.

Winds do not blow directly from high- to low-pressure areas. As a result of the rotation of the earth they are deflected. The relationship between winds and pressure, known as the *Coriolis effect*, was first recognized by the French mathematician, Coriolis (1792-1843). It influences any freely moving fluids such as air and water. Consequently any object or fluid moving horizontally in relation to the earth's surface tends to be deflected —because of earth's rotation—to the right of its path of motion in the northern hemisphere and to the left in the southern hemisphere. This deflection sets up clockwise and counter-clockwise circulations around the cellular pressure systems. The Coriolis effect is, of course, absent at the equator but increases progressively polewards. In addition to the Coriolis effect and the pressure gradient, friction is another factor that affects the direction and velocity of air movement. When moving air comes in contact with the surface of the earth its velocity is reduced, the extent depending on how rough the surface is.

In general the atmospheric circulation near the surface of the earth can be explained as follows. Heated air in the equatorial region rises and flows polewards. As it moves polewards it cools rapidly and is also deflected by the earth's rotation. These two factors lead the air to descend at approximately 30° latitude. Upon reaching the earth's surface some of this air moves towards the equator to complete a circulation system known as the *Hadley cell*, while the rest moves polewards. This air moving polewards is also deflected to the east (westerlies). Where this warm air comes in contact with the much cooler polar easterlies the warmer air is forced to rise and cells and troughs of low pressure develop. This area of contact is one of the most turbulent weather zones on earth. A description of what happens here is included in the section 'Air Masses, Fronts, and Storms'.

It is important to remember that Figs. 3-17 and 3-18 represent average monthly conditions and that a daily weather map for these months would probably look substantially different. One part of the world pressure system that changes the least from day to day is the Hadley cell. The trade winds that form this cell are among the most consistent winds on the earth's surface, partly because of the more uniform temperatures in the lower latitudes. The westerlies and polar easterlies, on the other hand, show great variation from day to day (see Fig. 3-21) and consequently are responsible for the changeable weather conditions characteristic of the middle latitudes.

MONSOON WINDS

From the foregoing analysis of the seasonal variations in winds and pressure, it is apparent that over large land masses there is a marked drop in average pressure between winter and summer. This in turn results in surface winds that flow from land to sea in winter and from sea to land in summer, causing fairly large seasonal differences in temperature and precipitation. Winds that reverse their direction between winter and summer are called monsoons. The name monsoon is derived from the Arabic word *mausim*, meaning season. Originally it referred to the winds that blow over the Arabian Sea for six months from the northeast and for six months from the southwest. Now its meaning has been enlarged to include certain other winds that blow with regularity at definite seasons of the year and particularly affect south and east Asia.

The cause of monsoons is more complicated than the explanation above might suggest. Although pressure changes caused by temperature changes may be sufficient in the middle latitudes to bring about a complete change of wind direction between winter and summer, winters are not severe enough in the low latitudes to permit the development of high pressure over land as a result of temperature changes. Instead, it is thought that the off-shore air movement in the winter, or period of low sun, is a result of the normal migration of the wind systems following the course of the sun. The tropical easterlies prevail during the period of low sun (winter), while in the period of high sun (summer) the tropical westerlies are the dominant winds.

UPPER ATMOSPHERIC CIRCULATION AND JET STREAMS

Up to now air movement in the lower part of the troposphere has been discussed. In the upper part of the troposphere pressure and wind characteristics are somewhat different. For example, the Coriolis effect and the pressure gradient force are exactly balanced. As a result, winds flow parallel to the isobars and form a circular flow around the cells of high and low pressure.

Two wind systems dominate the upper air flow at altitudes between 9000 m and 12 000 m. First, the *upper-air westerlies* make a complete circuit of the globe in both hemispheres between approximately 25° latitude and the pole. The second, called *tropical easterlies*, also make a complete circuit of the globe in the equatorial region. Between these two wind systems is a kind of upper air front that is frequently disturbed by the development of waves known as *Rossby Waves*. As shown in Fig. 3-19 these waves grow until they become so large that they are cut off, much like the formation of meanders on a river (p. 49). As a result of the development of these waves, warm air is carried polewards and cold air towards the equator. This type of heat exchange is part of the process that lets the earth maintain a heat balance (p. 10).

Jet streams are concentrated air flows that occur within the Rossby Waves. Consisting of one or as many as three separate jet cores, the winds in the centre of these cores often attain velocities of 300 to 450 km/h. The most important consequence of jet streams is their influence on weather at the earth's surface. It is generally believed that when the Rossby Waves have developed to their fullest, the exchange of polar air moving towards the equator and tropical air polewards is at its maximum. This would also appear to be a time of storms (e.g. middle latitude cyclonic storms—discussed in the next section) and generally changeable weather. When the waves are small, weather conditions are more stable.

One example of the influence of the Rossby Waves and the jet stream occurred in the unusual winter of 1977. During a large part of this winter the jet stream swung much further south than normal, extending over the southeastern United States. This development affected weather over a large part of North America. Record warm temperatures and drought occurred in the western part of the continent, causing water rationing, forest fires, and crop losses. In the northeastern states and adjoining parts of Canada unusually low temperatures with high snowfall resulted in energy shortages and a breakdown of transportation services in many areas. Thus, while the upper-air winds and the jet streams are not yet completely understood, there is no doubt that they have an important bearing on our weather conditions.

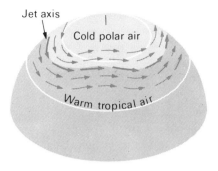

A. Jet stream begins to undulate

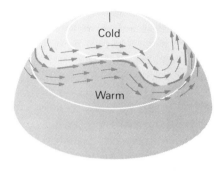

B. Waves begin to form

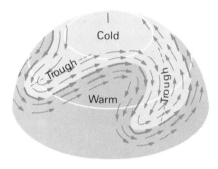

C. Waves strongly developed

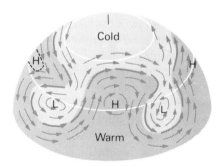

D. Cells of cold and warm air are formed

Fig. 3-19. Rossby waves and the jet streams. The heavier arrows show the position of the jet stream.

1 Make a cross-section diagram along a meridian of longitude (a semi-circle) from pole to pole. Identify the predominant winds, pressure, and circulation cells (see Figs. 3-17 and 3-18).
2 Referring to Figs. 3-17 and 3-18, find those areas of the world influenced by monsoon winds. Which of these seem to be a result of changes in pressure caused by temperature change? Which are better explained by the latitudinal shift of the wind system?

What large land masses do not experience monsoon winds? Explain.
3 What are some of the differences in temperature and precipitation that a place with monsoon winds might experience?

AIR MASSES, FRONTS, AND STORMS

Air Masses

Air that has assumed temperature and humidity in one area will retain those conditions as it moves to another area, and thus may bring about a change of weather. The large bodies of air involved in such transfers of heat and moisture are known as *air masses*. For example, both the British Isles and the Labrador coast of Canada are located at the same latitude, yet experience very different weather and climate. The British Isles are almost always under the influence of an air mass whose heat and humidity have been established over the Atlantic Ocean, while the Labrador coast is influenced by air that has moved across the North American land mass.

Air masses can only originate over either a large land mass or a large body of water. They develop where the air stays for a time on either land or water, or where it flows for long distances across land or water (Fig. 3-19). In either case the lower layers of air develop relatively uniform heat and moisture characteristics that vary from hot to cold and from dry to moist. The area where this occurs is known as the *source region* of an air mass. As the air mass moves away from its source region, travelling along in the general circulation of the atmosphere, it carries with it the temperature and humidity acquired over its source region. Its temperature and humidity influence the weather and climate of the places it passes over. Naturally the farther the air moves from its source region, the more its original temperature and humidity are modified.

Air masses are usually associated with areas of high pressure where the air is subsiding, but they can develop wherever they are in contact with a part of the earth's surface for a period of days. Low-pressure areas are unlikely source regions for air masses because, in such areas, bodies of air with different temperatures and humidity characteristics are converging.

A CLASSIFICATION OF AIR MASSES

In order to explain the climate of most areas of the world, it is necessary to understand the character and movement of air masses (Fig. 3-20) and how they influence different regions.

Air masses are classified according to their source region. Source regions are found in the low latitudes and in the high latitudes. Thus air masses are designated by the capital letters T (tropical) and P (polar) respectively. The latitude is important in determining the temperature of an air mass. Each air mass is also classified according to whether the source region is over land or over water. The lower case letters c (continental) and m (maritime) are used to indicate this difference. The moisture content of the air mass is determined by whether it originates over land or water. A detailed classification is shown in the chart below.

	Air Mass	Source Region
1	Maritime Polar	Oceans poleward of 40° latitude.
2	Continental Polar	Continents poleward of 60° latitude.
3	Maritime Tropical	Sub-tropical high-pressure areas over the oceans.
4	Continental Tropical	Low-latitude land areas, chiefly deserts.
5	Arctic (Antarctic) A (AA)	Surfaces of snow or ice.
6	Maritime Equatorial ME	Oceans close to the equator.

STUDY 3-10

1 Describe the characteristics of each of the world's major air masses at their sources in summer and in winter.

2 What air masses influence the following areas of North America (Fig. 3-20)? Describe the temperature and humidity brought to

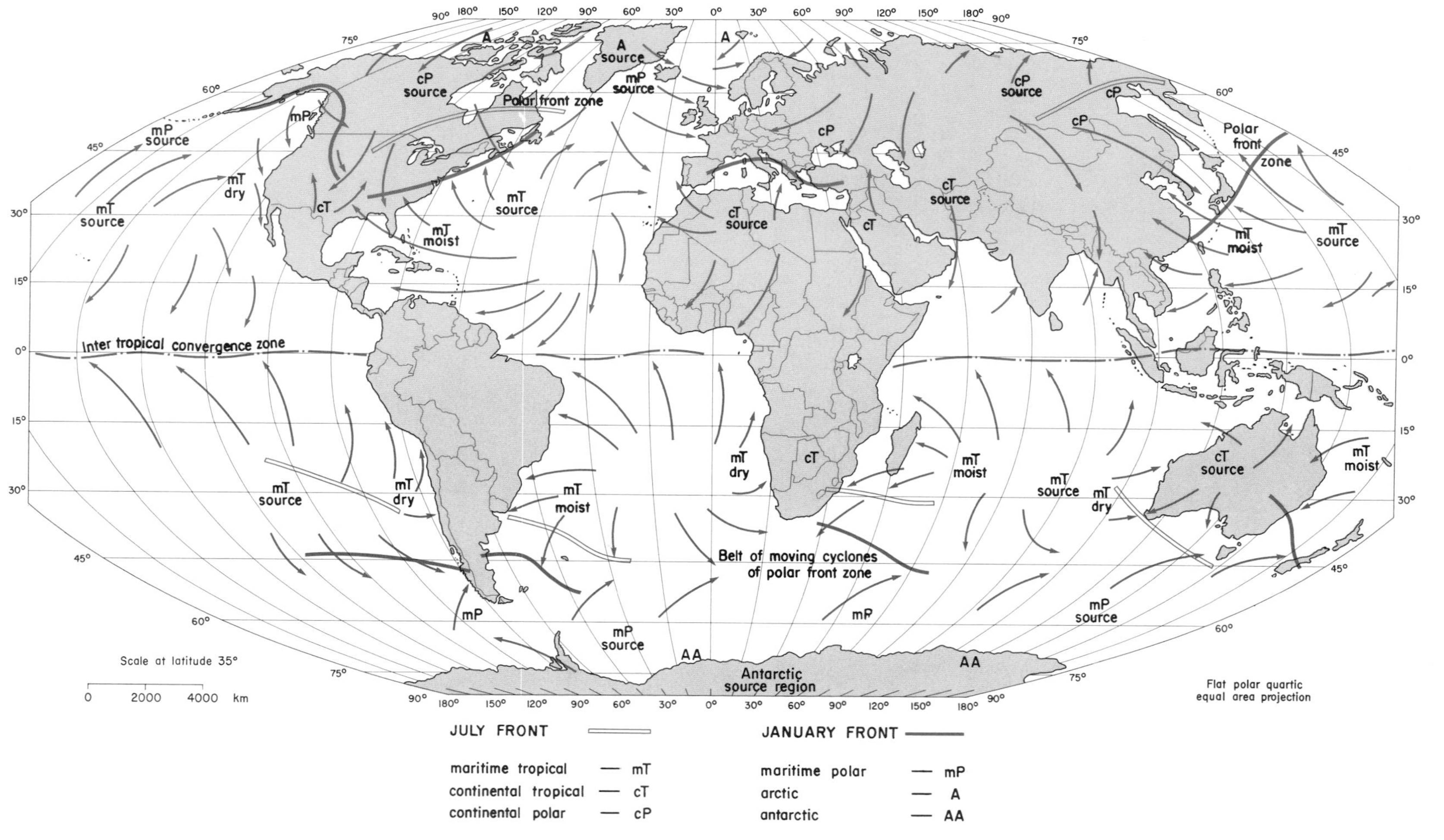

Fig. 3-20. Principal air masses of the world.

JULY FRONT —

JANUARY FRONT —

maritime tropical	— mT	maritime polar	— mP
continental tropical	— cT	arctic	— A
continental polar	— cP	antarctic	— AA

Scale at latitude 35°

0   2000   4000   km

Flat polar quartic
equal area projection

these areas by air masses in summer and in winter:

    a the Gulf Coast of the United States;

    b western British Columbia;

    c the New England states;

    d the lower Great Lakes area.

3 Some air masses will normally be more stable than others. For example, the air mass influencing the coast of British Columbia in summer will have been cooled at its lower levels by the ocean and, because of its relatively small lapse rate, will be fairly stable. Comment on the stability of the air masses influencing the Gulf Coast of the United States and the New England states. What will happen to the stability of the air masses that influence the Gulf Coast in the summer as they move inland?

4 Where would air masses with dissimilar characteristics meet, and what would occur in these areas when they did so? In what seasons of the year would this most likely occur?

Fronts

We have already examined how the general circulation of the atmosphere leads to the formation of areas of high pressure in the polar and sub-tropical latitudes where many air masses originate. There are also areas of low pressure in the tropics and the middle latitudes. Air masses with different temperature and humidity characteristics have dissimilar densities. When they come together in these regions of convergence (low pressure areas), they do not mix but remain separated by a sloping boundary called a frontal surface or *front*. On a weather map (Fig. 3-21) the front is indicated by a line of symbols that represents its location on the ground. However, remember that fronts are three-dimensional zones, that is, they vary in breadth from a few kilometres to 80 km and extend vertically as well as horizontally.

The position of a front is constantly changing. Usually one of the air masses separated by the front is more active and will be pushing the other back. When warm air is the stronger force, the front is referred to as a *warm front*, and conversely when cold air is the stronger, it is a *cold front*. Warm and cold fronts cause different weather conditions. A cold front occurs when a cold air mass enters a warm air zone and the denser cold air stays in contact with the ground. The warm air is forced to rise along a front sloping at approximately 1 in 40. Cold fronts may cause severe weather disturbances such as thunderstorms as the cold air, acting like a sharp wedge, causes the air in front of it to rise. Warm fronts occur where a warm air mass moves into areas of colder air. The warm air is again forced to rise but this time the slope is much less, ranging between 1 in 100 and 1 in 200. Generally along warm fronts weather conditions are more stable although clouds and precipitation will extend over a much greater area than along a cold front.

Because of the greater differences that normally exist between its converging air masses, the middle-latitude front (polar front) is of much greater importance than the tropical front (intertropical convergence zone). In the middle-latitude areas of the world the thrusts of warm or cold air—and the storms associated with these rapidly shifting air masses—cause the highly changeable weather that characterizes the climate of these areas.

STUDY 3-11

1 Referring to Fig. 3-20, indicate the principal air masses in summer and winter on outline maps of North America for each of these seasons. Show the source regions and movements of these air masses and also show the typical position of the polar front in each season.

    a Why does the position of the polar front shift?

    b What air masses does this front normally separate?

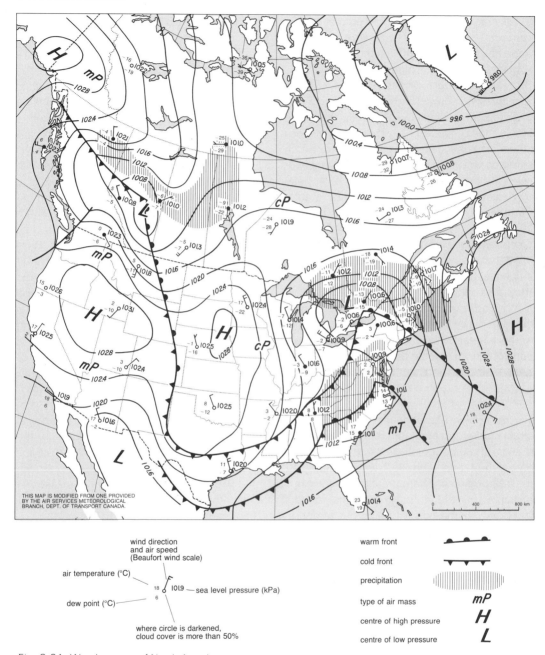

Fig. 3-21. Weather map of North America.

c Why is the forecasting of weather of great importance in the latitudes influenced by the polar front?

2 Compare the winter map of air masses with the weather map (Fig. 3-21). Do the position of the front and the characteristics of the air masses conform to the general conclusions you have already reached in Question one?

Middle-Latitude Cyclones and Anticyclones

To understand the nature of the contact between unlike air masses in the middle latitudes, we must look at the daily weather conditions. A daily pressure map (Fig. 3-21) provides a more realistic representation of atmospheric conditions than a map that gives only the average conditions over a whole month. It demonstrates how important aspects of weather and climate can be obscured on maps of average conditions. On the daily map, a series of low-pressure and high-pressure cells can be seen together with the fronts. These cells, which dominate the weather map, are called middle-latitude cyclones and anticyclones. Our understanding of them has developed rapidly since the First World War and has resulted in more accurate weather forecasting.

Their existence was first recognized by the Norwegian-born American meteorologist Jakob Bjerknes during the First World War. He used the word 'front' partly because of the similarity between this atmospheric phenomenon and the fighting fronts that existed in France. Just as the armies met and moved back and forth along a sharply defined line, so masses of cold polar air meet the warm, moist air from the south. Instead of mixing, these unlike air masses remain separated along a front. The front is constantly moving back and forth, and widespread cyclonic storms may form along it.

Middle-latitude cyclones are eastward-moving centres of low pressure. In some ways they resemble whirlpools that form between adjacent currents of water moving at different speeds. The convergence of air masses towards such centres produces cloudiness and precipitation. The cyclones are carried in the westerly wind system along the polar front.

Anticyclones, on the other hand, are centres of higher pressure that move along in the same way and direction as the cyclones (although they are not associated with fronts). It is the subsidence and outward spreading of air away from this type of centre that produces clear, sunny weather. The surges of cold polar air (cP) towards the equator are associated with anticyclones, particularly the northerly flow on the eastern side of the anticyclone. An example of a relatively mild outbreak is shown on Fig. 3-21.

Figure 3-22 shows the development of a cyclonic storm, while Fig. 3-21 shows a mature cyclonic storm over eastern North America. Several anticyclones can also be seen on the same map.

STUDY 3-12

1 With the help of the information provided by Figs. 3-21 and 3-22, describe the general characteristics of a middle-latitude cyclonic storm using the following headings:
   a areas of occurrence;
   b speed and direction of movement;
   c air masses involved and their characteristics;
   d the area of land that might receive precipitation from such a storm.

2 The cyclonic storm is a wave-like disturbance along a front. It is the horizontal convergence of air along a front (forming, in fact, two fronts) that normally triggers precipitation. Why are two fronts formed? How does the precipitation differ along them?

3 The cold front is moving more quickly than

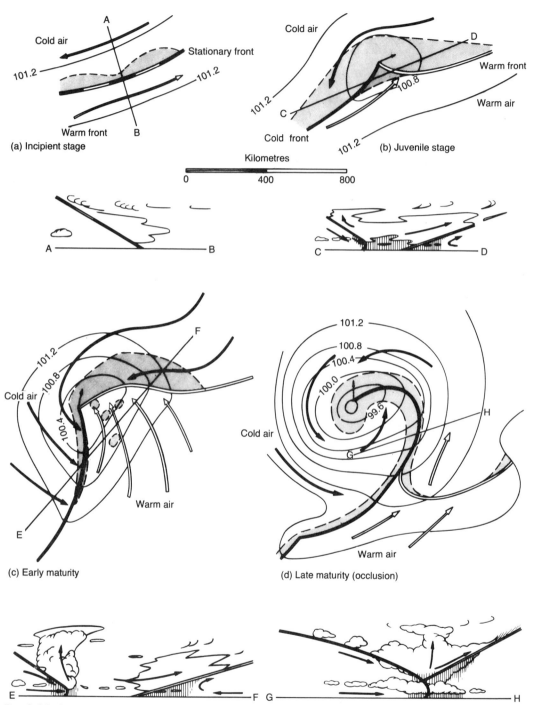

Fig. 3-22. Stages in the development of a mid-latitude cyclonic storm. The four stages are approximately 24-36 hours apart, while the distance the storm travels from (a) to (d) might be several thousand kilometres. The cross-sections show characteristic cloud formations and areas of precipitation.

the warm front and eventually overtakes it. Explain what happens to the cyclonic storm when this occurs. What stage does this represent?

4 Cyclonic storms are more severe in winter than in summer. Why is this normally the case? The path followed by these storms varies with the seasons. Explain this.

5 The passage of a well-developed cyclone and the anticyclone that often follows brings about changes in the weather that, in their main features, are usually quite predictable. Examine carefully Figs. 3-21 and 3-22 and write an account of these changes. Discuss changes in pressure, wind direction, temperature, and precipitation. Imagine that you are located at a point in advance of the storm and that it crosses your location from west to east. Describe what happens using the following headings:

    a the approach of the warm front;
    b between the warm and cold fronts;
    c after the cold front.

Tropical Disturbances

A number of weak, low-pressure areas occur in the intertropical convergence zone. Sometimes these appear on the weather map as waves in the isobars, known as *waves in the easterlies*. They occur frequently in the zone 5-30° latitude over oceans both north and south of the equator, but not over the equator itself. These waves are really feeble troughs of low pressure travelling *westwards* some 300-500 km/d. Although tropical weather is generally undisturbed, these waves affect the climate of the zone of trade winds, bringing thunderstorms and scattered showers. Similar to these waves are weak tropical lows. Although their pressure gradients are weak and their wind systems poorly developed, they are an important source of clouds and rainfall.

Tropical cyclones are called *hurricanes* in the Caribbean and *typhoons* off the China coast, and are similar in their basic characteristics to middle-latitude cyclones. They occur less frequently than middle-latitude cyclones but are much more catastrophic in their effect. Tropical cyclones develop only over water, usually at temperatures over 26°C, and normally between 8° and 15°N and S. At least some originate from a wave in the easterlies or a weak tropical low that intensifies, growing into a deep, circular low-pressure area (95.0 kPa or lower at the centre), with a diameter of anywhere from 150-500 km (see Fig. 3-24). Although the specific mechanisms that trigger hurricanes are not properly understood, we do know the conditions that generate these storms. Since the storm originates over warm tropical waters, the amount of water vapour in the atmosphere will be high. When precipitation begins in a rising air mass it will result in the condensation of large volumes of water vapour with the accompanying release of a large amount of latent heat. This release causes the adiabatic cooling rate to decline, thereby increasing the instability of the air mass relative to the air surrounding it. This increases the rate at which the air is rising, reducing the atmospheric pressure in the immediate area. Owing to this inequality of pressure, air moves quickly from the surrounding area into the pressure 'hole'. This extremely rapid air movement, together with the circular pattern of clouds and intense rainfall, are the principal characteristics of a hurricane. Since it is the large amount of water being condensed at high temperatures that is the driving force of the hurricane, it follows that the tropical storm dissipates as it travels inland or polewards.

The centre of a hurricane consists of a relatively cloudless, rainless zone 8 to 40 km wide, called the central eye. Here air is descending from high altitudes and is being warmed adiabatically. Elsewhere in the disc-shaped low, winds are spiralling at velocities of at least 120 km/h, and around the edge of the eye they can exceed 200 km/h. Once formed, these storms move westwards and then polewards at speeds of 10-15 km/h in the lower latitudes and 30-60 km/h farther polewards where their intensity begins to lessen. Eventually some may be transformed into mid-latitude cyclonic storms. Hurricanes occur at specific seasons, generally in the respective late summer season in each hemisphere when the equatorial low has shifted polewards.

The importance of tropical cyclones lies in their tremendous destructiveness, especially along low-lying coastal regions. Probably the most catastrophic typhoon of this century struck Bangladesh in November 1970. Winds of 160 to 240 km/h occurred together with high tides, producing a *storm surge* that inundated nearly the entire coastal region with depths of water in the delta and coastal region ranging between 60 cm and 5 m. Although advance warnings were broadcast, few people left the region; official estimates placed the deaths at around 400 000, most occurring from drowning. The losses and damage to crops, fields, livestock, and buildings were enormous.

Tornados

The tornado is a small, intense cyclone that appears as a black funnel cloud hanging from cumulonimbus clouds. On the ground it covers a small area, 90-500 m in diameter; its colour is a result of the moisture and debris being swept up by winds as high as 400 km/h. The exact triggering mechanism for tornados is also not properly understood. However they are most commonly associated with the conditions produced by the meeting of cold dry air and hot moist air. Thus they are associated with cold fronts, originate where turbulence along this front is greatest, and move at speeds of about 30 km/h. Most occur in the United States and are concentrated in the central states, particularly Oklahoma and Kansas.

Like the cyclone, the importance of the tornado lies in its destructiveness. Within the narrow limits of its path, destruction occurs quickly and is complete. Tornados often occur in groups—for example on 4 April 1974, 148 separate ones killed over 300 people and caused hundreds of millions of dollars damage in the central United States.

STUDY 3-13

Figure 3-24 shows a hurricane as it appears on a weather map and a barograph of the readings taken at one place as the disturbance passes. Organize the answers to the following questions into an explanatory note on tropical cyclones.

1 In what ways do tropical cyclones differ from middle-latitude cyclones? Mention the following:

  a the precise shape of the isobars, the steepness of the pressure gradients, the strength of the winds;

  b the presence of fronts and the character of the precipitation and temperature within each disturbance;

  c the direction of the winds;

  d the seasons of occurrence and their frequency;

  e the size of most tropical cyclones.

2 In what areas do tropical cyclones occur and what land masses do they affect? What are some of the protective measures that have been adopted by man in these areas against the strong winds and storm surges associated with these storms?

Thunderstorms

A thunderstorm (Fig. 3-8) is an intense but generally localized storm of relatively short duration. It is associated with very strong updrafts and downdrafts of air. When several different vertical air currents are involved, turbulence is fairly obvious in the thick cumulonimbus clouds that extend from 3000 to

Fig. 3-23. A photograph of Hurricane Gladys about 240 km southwest of Tampa, Florida, taken from the Apollo 7 Spacecraft. Moving cyclones such as this hurricane, as well as the mid-latitude varieties, contribute to the poleward transfer of heat.

15 000 m into the atmosphere. At the top the clouds spread out, and the resulting anvil-shaped formation is known as a thunderhead. Heavy precipitation (sometimes hail), gusting winds, lightning, and thunder are associated with these storms.

Some thunderstorms occur as a result of rising cells of warm moist air (convection). They may also develop along a cold front where warm, moist air is being undercut by cold air or when warm, moist unstable air is forced to rise over a landform barrier. Their intensity is related to the amount of latent energy available and its rate of release. Lightning, a huge electric arc or spark, is a prod-

uct of complex processes. In general it begins with a flash bringing down negative charges from the lower clouds and positive charges moving upwards from the ground. Charges occur within the clouds until their supply of negative charge is temporarily exhausted. Thunder is a result of the violent heating and expansion of air in the path of the lightning flash. This expansion causes the intense sound waves that we recognize as a thunder clap.

STUDY 3-14

1 High surface temperatures, instability, and

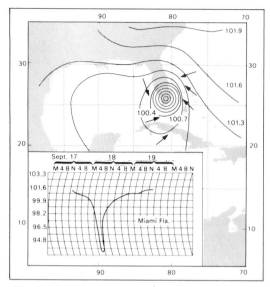

Fig. 3-24. A West Indies' hurricane as recorded at Miami, Florida. (Pressure is shown in kilopascals.)

humidity are the main conditions favouring the development of thunderstorms. Explain why all these conditions must be present.

2 Describe the distribution of thunderstorm activity in the world using the following headings:

    a latitude;
    b season;
    c time of day.

WORLD DISTRIBUTION OF TEMPERATURE AND PRECIPITATION

Temperature and precipitation are the most important elements of climate. In the previous sections of this chapter the factors causing variations in temperature and precipitation were analysed. Now it remains to examine the world patterns of temperature and precipitation and review the factors responsible for the differences that occur from place to place on the earth.

For a generalized study of world-wide variations in climate, average (mean) temperature and precipitation statistics are used. These are compiled, where possible, over 30 years. In more detailed studies these are only a beginning point, and a range of other statistics and methods of handling them become important.

World Distribution of Temperature

Average temperature statistics are derived from the mean daily temperature, which is an average of the highest and lowest temperatures during a 24 h period. In most places the lowest temperature occurs just before sunrise while the highest is about two to three hours after noon. Three other average temperatures are used frequently:
1 the diurnal or daily temperature range, which is the difference between the highest and the lowest temperatures during a 24 h period;
2 the mean monthly temperature, which is the average of the daily means for the month;
3 the annual temperature range, which is the difference between the hottest and coldest mean monthly temperatures.

The most useful statistics for an initial investigation of temperature are the averages for the hottest and coldest months, which for most places are January and July. Lines joining places having the same temperature, as shown on Figs. 3-25 and 3-26, are called *isotherms*. These maps provide the basis for an analysis of the general features and seasonal variations in the world's pattern of temperatures.

STUDY 3-15

In light of the examination of temperature that has so far been made in this chapter, make a careful analysis of Figs. 3-25 and 3-

26. The following questions can be used for guidance. The first section is concerned with features of world temperature common to all seasons, and the second concentrates on the variations in temperature from summer to winter. Give reasons for the answers to all these questions.

FEATURES COMMON TO THE MONTHLY TEMPERATURE MAPS

1 In what general direction do the isotherms trend, and in what direction does temperature change most rapidly? What causes this change?

2 a If the earth were either all sea or all land, what would be the relationship between isotherms and parallels of latitude?

b Where on the earth's surface do the isotherms most closely follow the parallels?

c Under what circumstances do the isotherms deviate most from their normal trend?

d Over what parts of the earth are located, in both January and July, (i) the hottest areas, (ii) the coldest areas? Explain these locations.

3 Contrast the rate of temperature change along the west coast of North America and along the west coast of Europe. What similar contrasts are there in other parts of the world?

SEASONAL VARIATIONS IN TEMPERATURE

Important seasonal variations in temperature can be observed on the January and July maps, Figs. 3-25 and 3-26.

1 Examine the 16°C isotherm over the northern hemisphere during January and July. On an outline map of North America mark the approximate position of this isotherm for January and July. Note how many degrees of latitude it moves through between these months (a) at the 135° west meridian, and (b) at the 105° west meridian. Using this example, explain the seasonal movement of isotherms.

2 Examine Fig. 2-13 and give three examples of warm and cold currents reducing or intensifying the seasonal bending of isotherms.

3 Annual temperature range can be used as an approach to seasonal variations in temperature.

a Find areas where the annual temperature range is over 30°C and explain why the range is so great in these particular areas.

b Why do the low latitudes generally experience a low annual temperature range?

c Name the areas outside the tropics that have a low annual temperature range (under 15°C), and explain why it is so low there.

World Distribution of Precipitation

The variation from place to place in the distribution of precipitation is one of the most important aspects of climate, perhaps even more important than the differences in temperature. Variations in precipitation directly or indirectly influence many other aspects of geography.

The most widely recognized climatic variation is in the annual amount of precipitation. Other important variations are seasonal distribution and reliability—the extent to which the total annual precipitation in any one year may be close to the statistical average for the place concerned.

STUDY 3-16

The pattern of annual precipitation around the world can now be explained as a result of the influences that have already been studied. Refer to Fig. 3-27 in answering the following questions. (Lines joining places having the same total annual precipitation are called *isohyets*.)

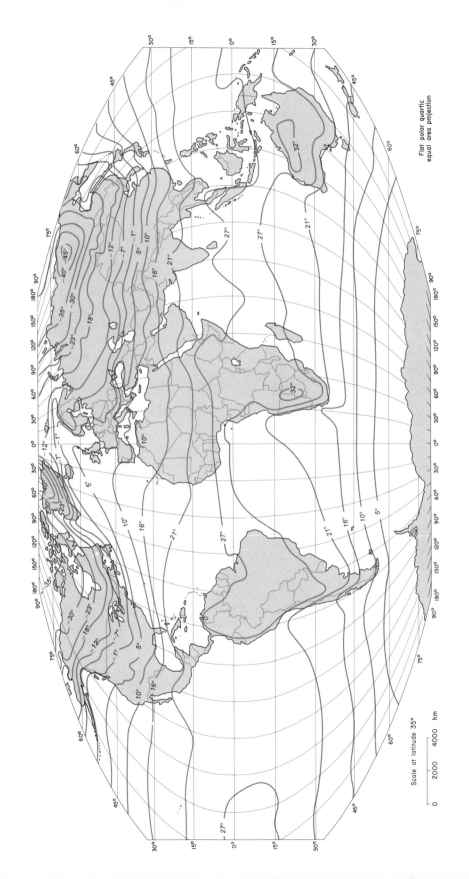

Fig. 3-25. Average sea level temperatures. January.

Flat polar quartic
equal area projection

Scale at latitude 35°

0   2000   4000   km

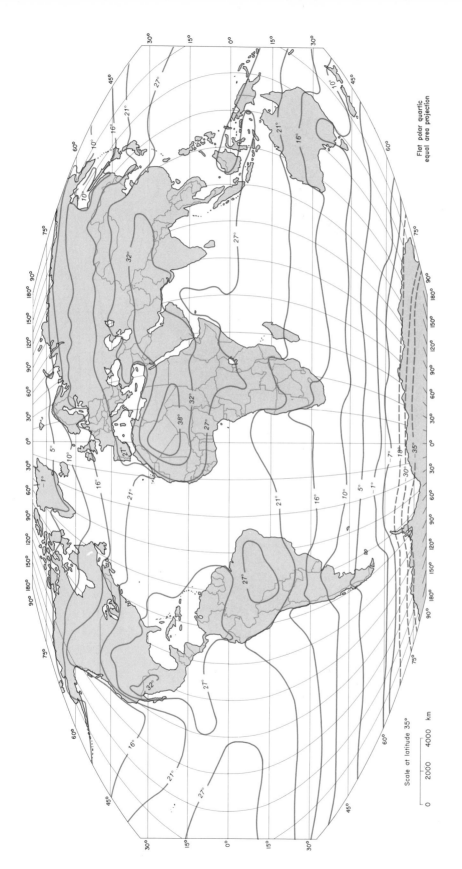

Fig. 3-26. Average sea level temperatures. July.

Flat polar quartic
equal area projection

Scale at latitude 35°

0   2000   4000   km

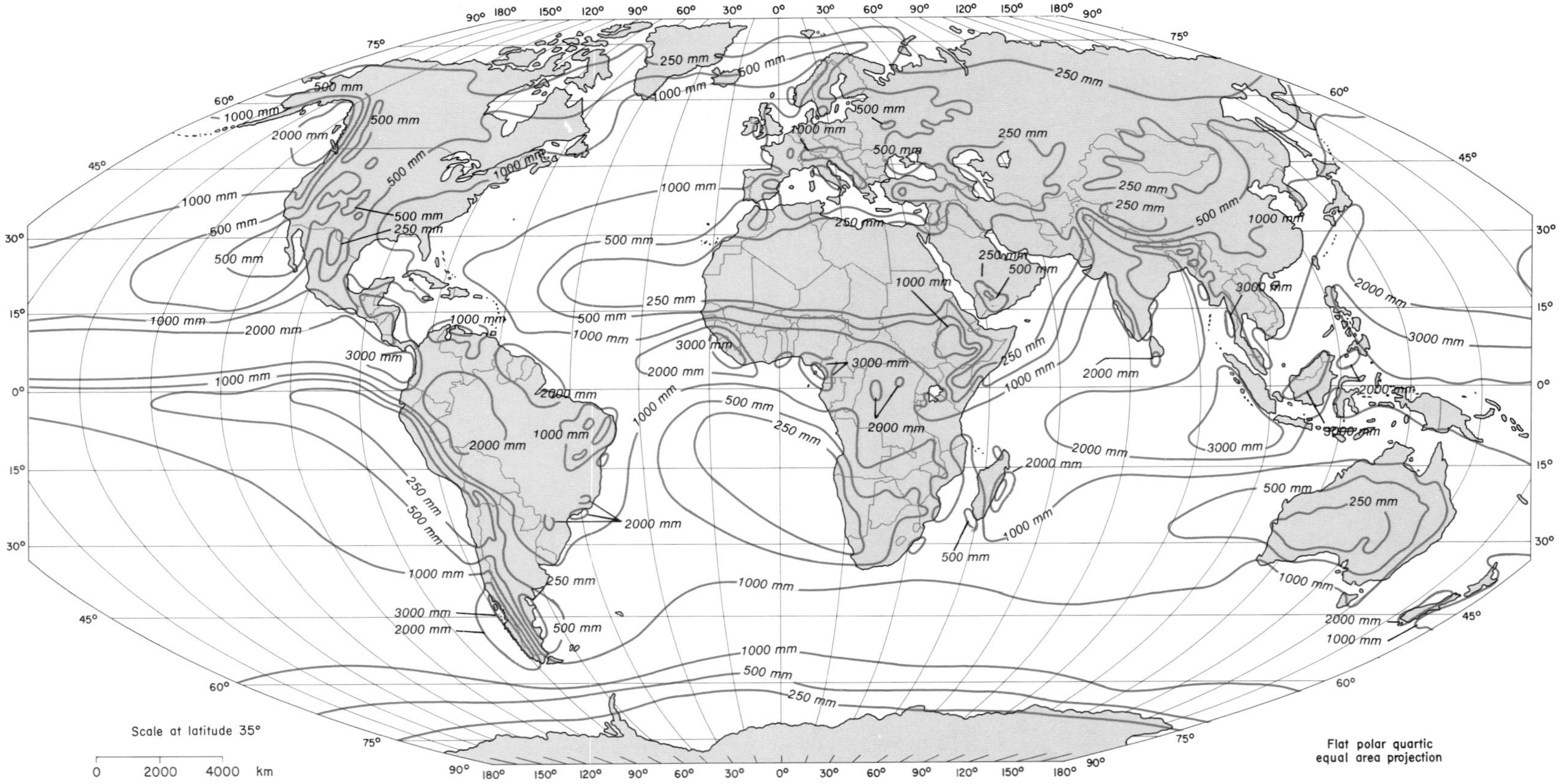

Fig. 3-27. Average annual precipitation.

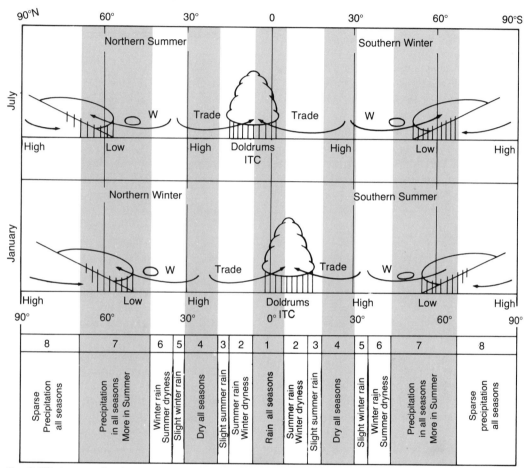

Fig. 3-28. A generalized cross-section of the atmosphere representing July in the top diagram and January in the bottom one. (Compare this with Figs. 3-17 and 3-18.) The zones of seasonal precipitation are indicated. It must be emphasized that this diagram is highly generalized, and there are many exceptions (after Petterson, *Introduction to Meterology*, McGraw-Hill, New York, 3rd edition, 1967.)

WORLD DISTRIBUTION OF MEAN ANNUAL PRECIPITATION

1 To establish a simple basis for describing the general pattern of precipitation distribution, make a number of meridional cross sections from pole to pole. Profiles taken along the west and east coasts of North America and the west coasts of Europe and Africa will quickly reveal a pattern. Important variations from this pattern can be seen in similar profiles across the land mass of Asia.

2 How much annual precipitation do most areas within the tropics receive? Relate the temperature characteristics of these areas to the amount of precipitation. What influence does the atmospheric pressure in tropical areas have on the vertical movement of air and on the amount of precipitation? What is the direction of the prevailing winds in the tropics? Which coast usually receives the highest precipitation? Give some exceptions to this and the reasons for them. What combination of circumstances would result in extremely high mean annual precipitation in tropical areas? Explain why daily maximums in temperature at the equator are not the highest on earth. Where are they the highest?

3 What generally happens to the amount of mean annual precipitation poleward from the tropics and, specifically, on the west coasts of continents? Give examples and specific statistics. What pressure systems are well developed on the eastern side of the oceans in these latitudes? Relate these to the pattern of precipitation. Do the ocean currents that parallel the coasts in these areas have any influence on the amount of precipitation?

4 Giving examples, describe the annual amount of precipitation in middle-latitude areas; treat the continental interiors as an exception. In what general direction are the winds blowing in these areas? What zonal pressure system influences the vertical movement of air? What daily weather features influence the amount of precipitation in these areas? Using North America and Eurasia as contrasting examples, describe and explain the west to east variation in precipitation over middle latitudes.

5 Explain the annual precipitation of high latitudes.

SEASONAL DISTRIBUTION OF PRECIPITATION

The seasonal lack (or excess) of precipitation is in some places more important than the annual amount. For instance, the availability of moisture in the soil during the growing season is particularly vital. Few areas of the world receive a truly even distribution of precipitation throughout the year, although many receive adequate precipitation at all seasons. Areas adjoining those parts of the earth that receive very high or very low precipitation often have an uneven distribution of precipitation. Such localities (for example the Mediterranean region and coastal and southern California) experience different amounts of precipitation in different seasons as a result of the seasonal movement of the pressure and wind systems (Fig. 3-28). In other areas, such as India, the seasonal variation in precipitation is brought about by a complete reversal of the winds resulting from seasonal pressure changes. The monsoon winds of Asia are the best example, although other continents experience less noticeable but sometimes important variations. The intense heat of continental interiors during the summer months also helps to bring about precipitation as a result of convection.

RELIABILITY OF PRECIPITATION

Annual amounts of precipitation vary from year to year. In some parts of the world this variation is small, and the reliability of precipitation is high. In others, reliability is low, which means that the total precipitation in any year may be substantially greater or less than the long-term average. Reliability may be

Fig. 3-29. Flooding controlled along the Mississippi River levee to protect adjacent farms. Fig. 3-30. The damage caused by uncontrolled flooding of the Red River in Winnipeg. Flooding is a serious natural hazard in many areas. In what kinds of locations are floods most common? What circumstances can produce floods? Why have they tended to become more severe in areas where the natural landscape has been greatly changed by people? Discuss the ways in which flood damage can be reduced (see question 4, Study 7-6, p.191).

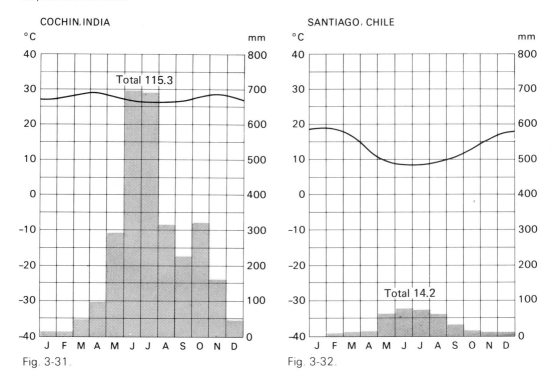

Fig. 3-31.

Fig. 3-32.

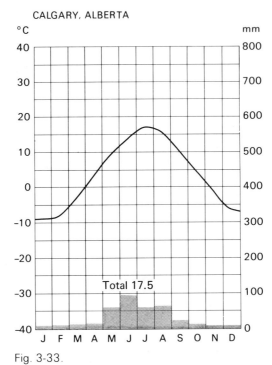

Fig. 3-33.

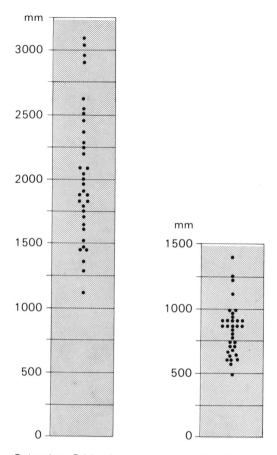

Colombo, Sri Lanka   Gibraltar

Fig. 3-34. Annual precipitation/dispersion diagrams. Each year's precipitation is shown by a dot on the vertical scale. In this way it is possible to see the range of wet years, average years, and dry years (after Monkhouse).

expressed as a percentage departure from the average and, as a general rule, decreases as the total annual precipitation decreases.

STUDY 3-17

1 What are the reasons for the uneven seasonal distribution of precipitation at each of the places shown in Figs. 3-31, 3-32, and 3-33?

2 In what areas of the world and under what conditions is reliability an important characteristic of precipitation?

3 Explain the method used to indicate reliability in the graphs for Colombo and Gibraltar (Fig. 3-34). Explain the difference in reliability of precipitation for these two stations.

4/ ASPECTS OF CLIMATE

We have discussed the conditions of the atmosphere and the world patterns of the two principal elements of climate: temperature and precipitation. In order to understand the way weather and climate vary from place to place, we should now examine the types of climate that presently exist over the earth's surface.

It is important to remember that climate types are not static. While change occurs slowly, the fact that climates have changed and will continue to do so is of importance to our future. In recent years scientists have made considerable progress in compiling a record of past climates. The object of these studies is to discover the forces responsible for climatic change. This information should make it possible to predict future climates.

THE CLASSIFICATION OF CLIMATE

It is clear that a great variety of different combinations of temperature and precipitation exist on the earth. Most of these combinations are found not just in one place; they occur in various parts of the world. It is possible to mark off vast areas covering large parts of continents that have similar temperature and precipitation characteristics, although within these areas there are many important local variations.

The criteria used to determine and draw boundaries between different climatic groups, and the size of the area chosen for study, vary depending on the purpose of the study. The most widely used classification of climate was first devised by Dr. Wladimir Köppen (1846-1940) in 1918 at the University of Graz in Austria. It is now a standard throughout almost the entire world. A modified version of his classification, shown in Figs. 4-1 and 4-2, has been adopted for this book. Köppen's classification has enjoyed widespread use because it is objective, quantitative, and can be universally applied, as long as statistics are available.

Köppen divided the world into five major climatic *groups* and assigned each a capital letter as well as a name. These groups are based on important characteristics of temperature and precipitation. A indicates the tropical humid climates (always hot and moist); B the dry ones; C the humid mesothermal ones (moist, with mild winters); D the humid microthermal ones (moist, with severe winters); and E the polar ones (always cold).

Each climatic group is subdivided into climatic *types* on the basis of the seasonal distribution of precipitation and/or the heat of the summer or the severity of the winter. Each is therefore a specific type of climate within its group.

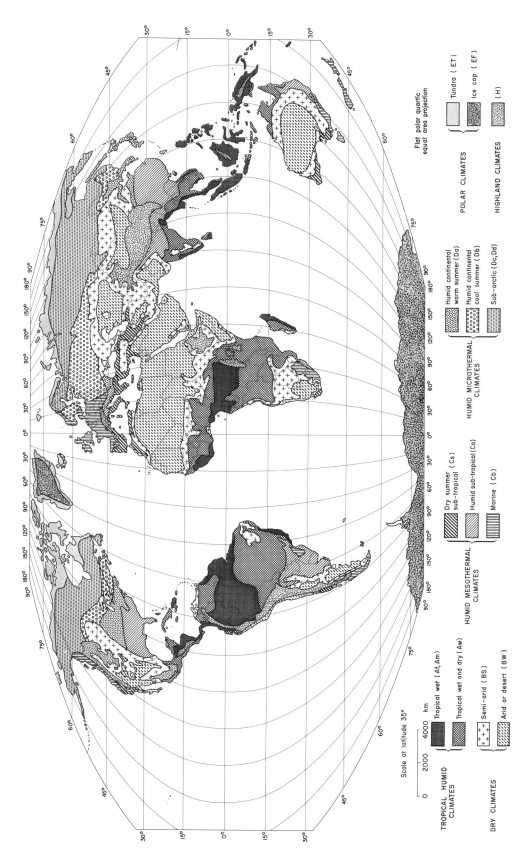

Fig. 4-1. Types of climate.

Fig. 4-2.

KÖPPEN'S CLASSIFICATION OF CLIMATES   Principal

Group	Secondary Symbols		Types (symbol and name)
A *Tropical Humid* (always hot and moist) Average temperature of coldest month 18°C or over	f (always moist at least 60 mm of precipitation every month)		Af *Tropical Wet*
	m (excessively moist during hot season with driest month less than 60 mm)		Am *Tropical Wet, Monsoon*
	w (at least one month less than 60 mm, longer dry season than m)		Aw *Tropical Wet and Dry*
B *Dry* Evaporation exceeds precipitation (see note below).	S (semi-arid)	h (hot-mean annual temperature over 18°C	BS *Semi-Arid*
		k (cool-mean annual temperature under 18°C	
	W (arid)	h	BW *Arid*
		k	
C *Humid Mesothermal* (mild winter, moist) Average temperature of coldest month between 18°C and − 3°C; at least one month above 10°C	f (always moist: at least 30 mm of precipitation every month)	a (warmest month over 22°C) b (warmest month less than 22°C, but at least 4 months over 10°C) c (1-3 months above 10°C)	Cfa, Cwa *Humid Sub-Tropical*
	s (summer drought at least one month below 30 mm)	a as above b	Csa, Csb *Dry Summer Sub-Tropical*
	w (winter dry season: wettest month at least 10 times as much precipitation as driest month)	a as above b	Cfb *Marine*

Group	Secondary Symbols		Types (symbol and name)
	f (as in C climates)	a as above b c d (temperature of coldest month less than − 38°C)	Dfa, Dwa *Humid Continental* *Warm Summer* Dfb, Dwb *Humid Continental* *Cool Summer*
D *Humid Microthermal* (severe winter, moist). Average temperature of coldest month under − 3°C; average tem- perature of warmest month over 10°C.	w (as in C climates)	a b c d	Dfc, Dfd, Dwc, Dw *Sub-Arctic*
E *Polar Climates* (always cold). Average temperature of warmest month under 10°C.	T (warmest month over 0°C) F (warmest month less than 0°C)		ET *Tundra* EF *Ice cap*
H *Highlands*			H *Highlands*

Note: B Climates

The following graphs can be used to determine whether the climate of a particular place is in the B category and whether it is BS or BW.

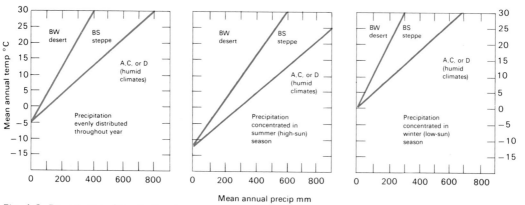

Fig. 4-3. Boundaries of the B climates.

PROCEDURE FOR STUDYING CLIMATIC GROUPS AND TYPES

The study of the five climatic groups and some of their main types requires the interpretation of various kinds of information. This text provides material that highlights some of the more important features of each of the climatic groups and types. These include all the maps, statistics, diagrams, and descriptions in this and the previous chapter. There are, of course, many other sources that can be consulted, some of which are included in the bibliography at the end of this book. Information supplied earlier in Chapter 3 will provide the theoretical basis for much of the reasoning required. The A or tropical humid climatic group is described and interpreted on pp. 95 to 98 and can be used as a model to complete the descriptions and explanations of the other four groups.

CHECK LIST

This check list is a guide to be used when describing and interpreting the climate of a locality or area. All the points on the list are not necessarily relevant to a specific climatic type. (Note that the numbered points under the Description sub-heading do not correspond to the same numbered points under the Explanation sub-heading.)

LOCATION

Area covered in relation to:

1 Latitude
2 Continental position
3 Adjacent types of climate
4 Landforms

TEMPERATURE

DESCRIPTION

1 Winter temperatures
2 Summer temperatures
3 Length and character of spring and fall
4 Mean and annual range of temperature
5 Diurnal range of temperature

The following relative terms, defined in relation to monthly temperatures, can be used. Summers over 25°C are very hot; between 20 to 25°C, hot; 10 to 20°C, cool; and under 10°C, cold. Winters between 10 to 20°C are warm; 0 to 10°C, mild; 0 to −10°C, cool; −10 to −20°C, cold; and less than −20°C, very cold.

EXPLANATION

1 Latitude
2 Altitude
3 The air masses influencing the area in both winter and summer

PRECIPITATION

DESCRIPTION

1 Mean annual precipitation (low, medium, high). Relate the effectiveness of precipitation to temperature.
2 Seasonal distribution of precipitation:
  a Evenly or unevenly distributed
  b Wet and dry seasons and amount in each
3 Form of precipitation
4 Reliability of precipitation

EXPLANATION

1 Landforms
2 The air masses influencing the area in both winter and summer
3 Stability of the lower atmosphere
4 Humidity
5 Storms

Steps to Follow

1 Examine all the information available on the group and type of climate being studied. Under which heading in the above check list might this information fit?

2 Make a preliminary examination of the location of the area in relation to the patterns of temperature (Figs. 3-25 and 3-26), precipitation (Fig. 3-27), winds and pressures (Figs. 3-17 and 3-18), and air masses (Fig. 3-20).

3 Referring to the check list, assemble detailed observations on relevant aspects of temperature and precipitation.

4 Referring again to the check list, give explanations for the aspects of temperature and precipitation observed.

5 In order to avoid missing important points, use the studies compiled for each climatic group and type being examined, and make a preliminary organization for the final written account.

6 In writing the final account, make certain that the essential aspects that distinguish a climatic group or type from all others are made clear.

TROPICAL HUMID CLIMATES / A

The tropical humid climates (A group) can be subdivided into three major types: tropical wet (Af), tropical wet and dry (Aw), and tropical monsoon (Am). The entire group is characterized by uniformly high temperatures. Consequently the three types are distinguished largely by variations in both annual and seasonal precipitation. The source material included provides temperature and precipitation statistics for two tropical wet stations (Af), and for comparison, a typical wet and dry station (Aw) and a typical monsoon station (Am).

The following analysis of the A climates has been written using the source material just mentioned, as well as the various maps, diagrams, and other written material in this and the previous chapter. In this account the main purpose is to distinguish the A climates from the other four groups. Where differences within the A group are especially important, they are noted and explained.

An Account of Tropical Humid Climates

The tropical humid climates are found in a somewhat interrupted belt, for the most part within 20° north and south of the equator (Fig. 4-1). The largest areas of A climates are on the continents of Africa and South America, the only land masses with a considerable east-west extent in these latitudes. Less extensive areas of A climates are found over parts of the Indian subcontinent, over parts of the mainland and islands of Australia, and over most of central America and the islands of the Caribbean.

For the most part, A climates are bounded by and merge with B climates, although in some localities, notably in Asia, they adjoin the humid sub-tropical climates.

GENERAL CHARACTERISTICS

A climates are the true tropical or winterless humid climates. They are distinguished from all other humid climates by their temperature. In accordance with Köppen's definition, no month has a mean temperature of less than 18°C. Combined with these consistently high temperatures is a generally high annual precipitation, with most areas receiving over 1500 mm.

Within this group there is considerable variation in the seasonal distribution of precipitation. On this basis, the group is subdivided into the Af, the Aw, and the Am types.

TEMPERATURE

Together with their consistently high temperatures most areas within the A group have mean annual temperature ranges of less than 4°C while almost all the A climates have annual temperature ranges of less than 6°C. The continually high temperatures are not surprising. The sun is never far from being directly overhead, and the amount of insolation is greater here than anywhere else in the world. As a result of the uniform length of day and night, seasonal variations in insolation are small. The moderating effects of maritime winds and air masses also tend to keep annual temperature ranges low.

Average daily temperature ranges for this group are usually between 6°C and 10°C. They are therefore usually higher than the

Source Material

Fig. 4-4.

CLIMATIC STATIONS:
In this table and the ones that follow in this chapter, temperatures (T) are given in degrees Celsius, and precipitation (P) is given in millimetres.

Georgetown (Guyana) Af

	J	F	M	A	M	J	J	A	S	O	N	D
T	26	26	26	27	27	27	27	27	28	28	27	26
P	203	114	175	140	290	302	254	175	81	76	155	287

Total 2252 mm

Beira (Mozambique) Aw

	J	F	M	A	M	J	J	A	S	O	N	D
T	28	28	27	26	23	21	21	21	23	26	26	27
P	277	213	256	107	56	33	30	28	20	132	135	234

Total 1521 mm

Bombay (India) Am

	J	F	M	A	M	J	J	A	S	O	N	D
T	24	24	26	28	29	29	27	27	27	28	27	26
P	2	2	2	<1	18	485	617	340	264	63	13	2

Total 1809 mm

MEAN TEMPERATURES (°C)

	COOLEST MONTH			WARMEST MONTH			
	Month	Daily max.	Daily min.	Month	Daily max.	Daily min.	Absolute extremes
Singapore	26	30	23	27	31	24	36 and 19

annual temperature range, and, contrasted with the oppressive heat during the day, the slightly cooler nights are welcome. The relatively high diurnal temperature ranges give rise to the statement that in the tropics 'night is the true winter'. However, owing to the excessive humidity, even the more bearable night temperatures are high compared with those in other parts of the world.

One of the most noteworthy aspects of tropical humid climates is the lack of variation in the pattern of temperatures from day to day, a circumstance that emphasizes the sun as the dominant factor in temperature control.

PRECIPITATION

Most areas within this group receive more than 1500 mm of precipitation annually, and places with over 3800 mm are not uncommon. Within the A group precipitation decreases as one moves polewards. The Aw and Am climates are distinguished from the Af ones mainly on the basis of their noticeable dry season. However, in the Af type, the monthly statistics for rainfall vary much more than those for temperature. Despite the high annual amounts, precipitation is seldom evenly distributed in the group.

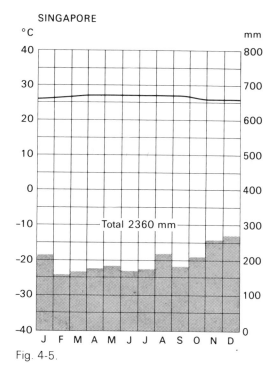

SINGAPORE

Fig. 4-5.

Essentially, the A climates coincide with the intertropical convergence zone. The northeast and southeast trades converge in a zone that moves northwards and then southwards with the apparent movement of the sun. Thus converging air masses combine

Fig. 4-6. The daily maximum and minimum temperatures for the months of January and July, for Singapore (after Trewartha, Robinson, and Hammond).

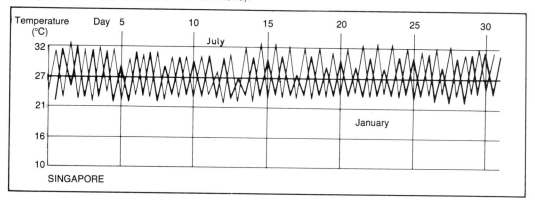

with the high temperatures and humidity to provide unstable conditions, ideal for the condensation of large quantities of water vapour. Much of the rain comes from cumulus clouds that are convectional in origin. The convectional currents are partly induced thermally, but they are also associated with converging air and weak tropical disturbances. Rainfall is often heaviest and most frequent in the late afternoon, but the 'waves in the easterlies' also cause precipitation during the night.

Owing to the seasonal movements of the intertropical convergence zone, the Aw areas are for part of the year under similar influences to those causing precipitation in the Af areas. For the other part of the year, they are under the influence of the drier stable air of the subtropical high-pressure systems. The stable descending air of the subtropical highs results in a distinct dry season of more than two months' duration in the Aw areas. In the Am areas, the dry reason is usually even more pronounced and the total precipitation is normally greater than the Aw type. The monsoon phenomenon responsible for this type is explained on page 68.

DRY CLIMATES / B

As their name suggests, the dry climates (B group) are distinguished from all others by their low precipitation. The effectiveness of precipitation, that is, the amount that is not evaporated and therefore available for plant growth, varies with temperature. Consequently no precise amount of precipitation can be used as a criterion to determine the boundaries of all B areas. In the high latitudes 500 mm of precipitation may be sufficient to maintain abundant moisture in the soil throughout the year, whereas the same amount falling in the hot tropics may result in a deficiency of soil moisture because of much higher rates of evaporation. Dry climates are therefore defined as those where potential evaporation exceeds precipitation. As a result

of rainfall deficiency, there is little surplus water to provide run-off. Thus few rivers of any size originate within such areas, though fairly large ones such as the Nile may cross them.

Dry climates are subdivided into semi-arid (BS) and arid (BW) by the formula that relates potential evaporation to precipitation (page 93). Each can be further subdivided into BWh or BWk, and BSh or BSk ('h' designating hot and 'k' designating cold). In this account dry climates are discussed as a group, but where important variations exist within the group, they should be noted.

One of the most outstanding characteristics of the B climates is the unreliability of the precipitation (p. 89). In some localities months and even years may pass without rain. When it does occur it is often torrential, forming underground water supplies that people in these regions depend on. Mean precipitation, therefore, has little significance in such regions. Some additional characteristics of this climate type are noted in the following brief description.

Suddenly, after hardly any twilight, the sun rises into the clear sky. In this dry atmosphere its rays are already scorching in the early morning, and under the influence of the reflection from rock and sand the layer of air next to the ground is heated rapidly. There is no active evaporation to moderate the rising temperature. After 9 o'clock the heat is intense, but goes on increasing till 3 or 4 in the afternoon, when the quivering mirage is sometimes seen. It cools slowly towards evening, and the sun, just before it sets, suffuses the cloudless sky with a glow of colour. In the transparent night the rocks and sand lose their heat almost as rapidly as they acquired it, and the calm of the atmosphere, which is so still that a flame burns without a tremor, also favours the cooling of the air. We shiver with cold and it is no uncommon thing in winter to find water on the surface of the ground frozen in the morning.[1]

Absolute extreme temperatures in or very close to the Sahara are:

	Absolute maximum	Absolute minimum
Biskra	49°C	− 1°C
Touggourt	50°C	− 3°C
Azizia	58°C	− 3°C
Tamanrasset	39°C	− 7°C
Bilma	47°C	− 2°C
Port Etienne	46°C	− 7°C

Because there is low humidity and a general absence of vegetation in most desert areas, wind velocities are often high. Consequently, life in these areas is dominated by wind and the sand it carries.

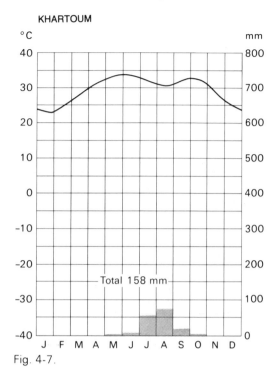

KHARTOUM

Total 158 mm

Fig. 4-7.

Fig. 4-8. The daily maximum and minimum temperatures for the months of January and July, for Yuma, Arizona (after Trewartha, Robinson, and Hammond).

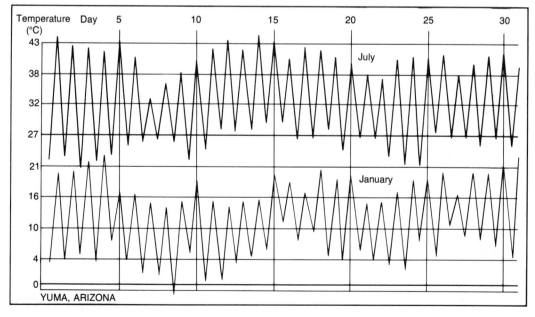

YUMA, ARIZONA

Source Material

Fig. 4-9.

CLIMATIC STATIONS:

Cairo (Egypt) BWh

	J	F	M	A	M	J	J	A	S	O	N	D
T	13	15	18	21	25	28	28	28	26	24	20	15
P	5	5	5	3	3	0	0	0	0	0	3	5

Total 29 mm

Medicine Hat (Alberta) BSk

	J	F	M	A	M	J	J	A	S	O	N	D
T	− 10	− 9	− 2	7	13	17	21	19	13	8	− 2	− 7
P	18	18	20	25	38	58	35	35	38	18	18	20

Total 341 mm

Lahore (Pakistan) BSh

	J	F	M	A	M	J	J	A	S	O	N	D
T	12	14	21	27	32	34	32	31	29	24	17	13
P	23	25	20	13	18	35	128	118	58	8	3	10

Total 459 mm

STUDY 4-1

1 Describe the location of the B climates in relation to pressure and winds and in relation to the land masses on which they are found. Both aspects of location are important influences on precipitation.

2 Explain how the low precipitation in coastal areas of low latitudes is influenced by the cool ocean currents that flow parallel to the coast.

3 The highest mean monthly temperatures as well as the highest daily temperatures in the world occur in B climates. Why is this surprising? Explain why it occurs.

4 What are the reasons for the generally high annual and diurnal ranges of temperature in the B climates?

5 Explain the reasons for variations in summer and winter temperatures within the group.

6 What differences are there between the patterns of weather in the areas of B climate in the low latitudes and those in the high latitudes?

HUMID MESOTHERMAL CLIMATES

The term *mesothermal* derives from the Greek *mesos*, meaning middle, and *thermos*, meaning hot. It refers to moderate temperatures. Unlike the tropical and the dry climates, the C group and the types within it are distinguished by both temperature and precipitation. This group does experience a definite winter, although nowhere is it severe. Precipitation is generally adequate, but there is considerable variation in both yearly totals and seasonal distribution. Since they are located in the middle latitudes, the C climates (and the D as well) are influenced by the convergence of unlike air masses, and thus changeability of weather is a notable characteristic.

There are three important types within the C group: the dry summer sub-tropical, also known as the Mediterranean climate (Csa, Csb), the humid sub-tropical (Cfa, Cwa), and the marine (Cfb).

STUDY 4-2

1 The Cs type of climate covers only about 2 per cent of the world's land area. Why then are its characteristics so widely known?
2 Note the location of this type in relation to the sub-tropical highs and the westerlies. Why is it regarded as an intermediate climatic type between the B group and the Cfb type?
3 What are two reasons for the mildness of the winter?
4 Though frosts are infrequent, they are important. Explain.

Dry Summer Sub-Tropical (Csa,Csb)

Source Material

Fig. 4-10.

CLIMATIC STATIONS:

Naples (Italy) Csa	J	F	M	A	M	J	J	A	S	O	N	D
T	8	9	11	14	18	21	24	24	22	16	13	10
P	93	70	73	73	50	38	18	23	73	133	113	118

Total 875 mm

San Francisco (California) Csb	J	F	M	A	M	J	J	A	S	O	N	D
T	10	12	13	13	14	15	15	15	17	16	14	11
P	115	92	74	37	15	3	0	1	5	22	51	109

Total 524 mm

MEAN TEMPERATURES (°C)

	JANUARY		JULY		Absolute extremes	
	Daily max.	Daily min.	Daily max.	Daily min.	low	high
Lisbon	13	8	27	18	− 2	39
Athens	12	6	32	22	− 7	43
Marseilles	11	3	28	16	− 13	38
Perth	29	17	17	9	1	44
Adelaide	30	16	15	7	0	48

5 Compare summer temperatures with those of the Cfb climate.

6 Note and account for the latitudinal variation in precipitation within many of the Cs regions.

No account of the American winter can omit a reference to the 'cold waves', which are defined officially as spells having a fall of temperature within 24 hours of at least a certain magnitude, down to, or below a specified figure, the limits being appropriate to region and season. In Florida the fall must be at least 10° and give a reading below 0° in winter, 2° in November and March; in New York the corresponding figures are 12°, − 7°, and − 2°. The strong cyclonic activity in the conti-

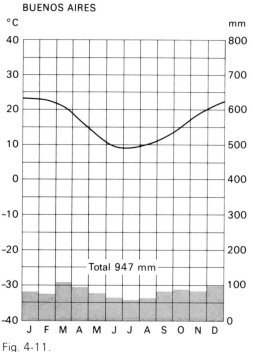

Fig. 4-11.

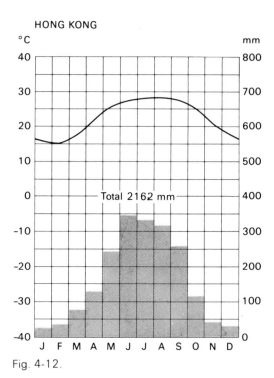

Fig. 4-12.

HUMID SUB-TROPICAL (Cfa, Cwa)

Source Material

Fig. 4-13.

CLIMATIC STATIONS:

Chungking (China) Cwa

	J	F	M	A	M	J	J	A	S	O	N	D
T	7	10	14	19	23	26	29	30	24	19	14	10
P	15	20	38	99	142	180	142	122	150	112	48	20

Total 1088 mm

Galveston (Texas) Cfa

	J	F	M	A	M	J	J	A	S	O	N	D
T	12	13	17	21	24	28	29	28	27	23	17	14
P	88	70	63	88	88	88	103	95	133	113	83	103

Total 1095 mm

nent has already been mentioned. As they travel east some pressure-systems become elongated in a north-south direction and polar air-masses sweep over great distances behind their cold fronts; many originate in the north-west where the clear skies of an anti-cyclone favour intense cold, and as they advance south they bring Canadian temperatures, modified somewhat by the journey, even to the Gulf of Mexico. On the average 3 or 4 cold waves a year reach the eastern States; they are an important factor everywhere in the variable weather for which the States are noted; the rapidity of the changes results from the rapid movement of the depressions.[2]

STUDY 4-3

1 Relate characteristics of the humid sub-tropical type to its location on the eastern side of continents.

2 What similarities in temperature but differ-

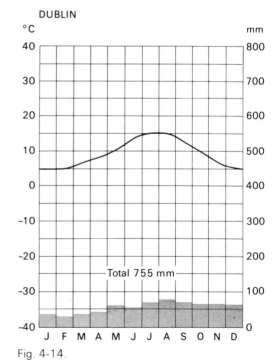

Fig. 4-14.

ences in precipitation exist between the humid sub-tropical climate and the Mediterranean climate?

3 'Sensible' temperatures refer to the sensations of temperature the body feels. For example, a hot humid day is less comfortable than a dry day with the same temperature since the loss of heat from the body is slowed down when the air is humid. A cold windy day feels colder than a day with the same temperature but little or no wind, because the loss of body heat is speeded up by greater evaporation. Thus the body, being sensitive to factors other than air temperature, will register impressions of heat or cold quite different from those suggested by thermometer readings. The high relative and high absolute humidities in summer can be related to *sensible* temperatures and to the generally low diurnal temperature ranges. Explain the relationships.

4 Referring to air masses, explain the variations in winter temperatures that occur within this type.

5 What influences do warm ocean currents have on this type?

6 Explain the reason for the direction in which rainfall decreases.

Marine Climate (Cfb)

Source Material

Fig. 4-15.

CLIMATIC STATIONS:

Dublin (Ireland) Cfb

	J	F	M	A	M	J	J	A	S	O	N	D
T	5	5	6	8	11	13	15	15	13	10	7	5
P	69	56	51	48	58	51	71	76	71	69	69	66

Total 755 mm

Hokitika (New Zealand) Cfb

	J	F	M	A	M	J	J	A	S	O	N	D
T	15	16	14	12	9	7	7	8	9	11	12	14
P	274	190	239	236	244	231	218	239	226	292	267	262

Total 2918 mm

SEASONAL DISTRIBUTION OF PRECIPITATION FOR MARINE STATIONS IN WESTERN EUROPE

	Percentages of the Yearly Total			
	WINTER	SPRING	SUMMER	AUTUMN
Bergen	30	18	20	32
Valentia	32	18	22	28
Brest	32	23	18	27
Paris	20	23	29	28

MEAN TEMPERATURES (°C)

	JANUARY			JULY				Ground frost mean annual
	Daily max.	Daily min.	Range	Daily max.	Daily min.	Range	Absolute extremes	number of days
Auckland, N.Z.	23	16	7	13	8	5	32 and 0.6	3

STUDY 4-4

1 Explain in detail how the Cfb type of climate can be located so far polewards and yet have a mild winter. Note and explain its west-coast location (compare Hokkaido and British Columbia) and its extent inland on different continents.

2 Compare average summer and winter monthly temperatures of marine climate stations with those recorded at interior locations at the same latitude.

3 Explain the great variation in annual precipitation from one Cfb region to another.

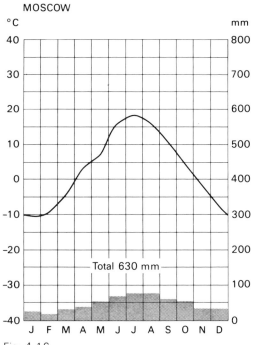

Fig. 4-16.

4 Explain fully the influence of middle-latitude cyclones as (a) triggers of precipitation, and (b) controls of weather. During what period of the year are they more common?

5 Explain why the diurnal range of temperature is so small, particularly in winter.

HUMID MICROTHERMAL CLIMATES / D

The term microthermal refers to low temperatures. It is derived from the Greek *mikros*, meaning small, and *thermos*, meaning hot. Microthermal climates are land-controlled and therefore have distinctly continental characteristics. Covering vast areas of North America and northern Asia, D group climates exhibit great variations in temperature. The southernmost type, the Dfa, Dwa (humid continental, warm summer), has mean monthly summer temperatures over 22°C and mean monthly winter temperatures in the − 7 to − 1°C range. This contrasts with the northernmost type, the Dfc and Dwc (subarctic), with summers in the 13 to 18°C range and winters from as low as − 51°C to as high as − 18°C. Despite these variations, this group has certain common characteristics:

1 The frost season is longer than that of the C group. Winters are colder and increase in severity towards the north. Snow remains on the ground throughout most of the winter. Of the two seasonal extremes, it is the winter cold rather than the summer heat that is most characteristic and distinctive.

2 Summers are warm for the latitude. Sum-

mers, and consequently the growing season, decrease in length towards the north.

3 Although precipitation is adequate, in many cases it is just barely so. The season of maximum precipitation is summer. Precipitation generally decreases from south to north.

Source Material

Fig. 4-17.

CLIMATIC STATIONS:

Montreal (Québec) Dfb

	J	F	M	A	M	J	J	A	S	O	N	D
T	− 10	− 9	− 3	6	13	19	21	20	16	9	2	− 7
P	83	81	78	72	72	85	89	77	82	78	85	89

Total 971 mm

Omaha (Nebraska) Dfa

	J	F	M	A	M	J	J	A	S	O	N	D
T	− 5	− 3	− 3	11	17	23	26	25	19	13	4	− 2
P	20	24	37	64	88	114	86	102	66	43	32	20

Total 696 mm

Okhotsk (USSR) Dwc

	J	F	M	A	M	J	J	A	S	O	N	D
T	− 24	− 22	− 12	− 7	0	6	11	12	8	− 3	− 15	− 21
P	2	2	5	10	23	41	56	66	61	25	5	2

Total 298 mm

STUDY 4-5

The source material above can be used to describe this climate group as a whole or the humid continental and sub-arctic types. In either case, the questions below will be a useful guide.

1 The severity of the D climates is largely a result of latitude and location on a continent. Comment on the influence of these factors.

2 Why does this climate group not exist in the southern hemisphere?

3 The effect of snow cover is to reduce winter temperatures even further, while at the same time protecting the soil from the very cold temperatures at the surface of the snow and also preventing deep freezing. Explain how the snow cover accomplishes this.

4 For most of this climate group summer is the season of maximum precipitation. Why

does more precipitation fall during the warmer months than the colder months? Why is this distribution fortunate?

TUNDRA, ICE CAP, AND HIGHLAND CLIMATES

Tundra climates (ET type) are found on the northern edge of the North American and Eurasian continents. These climates are distinguished from the sub-arctic type by their cooler and very short summers.

The ice cap climates (EF type) are found over the great, permanent, continental ice sheets of Greenland and Antarctica. They also include large areas of floating sea ice, particularly in the Arctic Ocean. This climate type has the lowest summer temperatures of any area on earth and, in the Antarctic, the lowest winter temperatures as well. Indeed, in no month does the average temperature rise above freezing.

Highland climates (H group) do not strictly constitute a group but rather are a category

Source Material

Fig. 4-18.

CLIMATIC STATIONS:

Sagastyr (USSR) (73°N 124°E) ET

	J	F	M	A	M	J	J	A	S	O	N	D
T	− 37	− 38	− 34	− 22	− 9	0	5	3	1	− 14	− 27	− 33
P	3	3	0	0	6	12	9	35	10	3	3	6

Total 90 mm

Quito (Ecuador) H alt. 2800 m

	J	F	M	A	M	J	J	A	S	O	N	D
T	12.5	12.8	12.5	12.5	12.6	12.8	12.7	12.7	12.8	12.6	12.4	12.6
P	80	98	120	175	115	38	28	55	65	98	100	90

Total 1062 mm

Pikes Peak (Colorado) H alt. 4233 m

	J	F	M	A	M	J	J	A	S	O	N	D
T	− 17	− 16	− 13	− 11	− 5	1	4	4	0	− 6	− 12	− 14
P	40	38	50	88	95	40	105	98	43	35	48	65

Total 745 mm

in which it is convenient to place mountains and plateaus in low and middle latitudes with an elevation greater than 1000 m. Variations in temperature and precipitation within this category are great. They are caused as much by differences in altitude as by variations in exposure to sun and wind. (Mountains and plateau regions of lesser elevation are included within the general climatic types characteristic of the surrounding lower lands.)

Temperatures (°C) at McMurdo Sound, Antarctica (4 years)[3]

| | MEAN | | ABSOLUTE | |
	Daily maximum	Daily minimum	Maximum	Minimum
J	− 2	− 8	4	− 16
F	− 7	− 13	1	− 23
M	− 13	− 19	− 3	− 29
A	− 19	− 26	− 7	− 41
M	− 21	− 29	− 8	− 46
J	− 21	− 30	− 6	− 44
J	− 21	− 31	− 9	− 47
A	− 22	− 31	− 8	− 46
S	− 21	− 31	− 9	− 51
O	− 17	− 24	− 4	− 41
N	− 7	− 14	1	− 22
D	− 2	− 8	− 6	− 16

STUDY 4-6

TUNDRA

1 Despite the low summer temperatures and the short growing season characteristic of this type of climate, some agriculture is possible in protected locations. What important factor partially compensates for the low temperatures and short growing season?

2 Explain why winters in the tundra are often not as cold as they are in the sub-arctic regions to the south.

3 It has been estimated that 75-90 per cent of the surface of the tundra is free of snow in the winter. Explain.

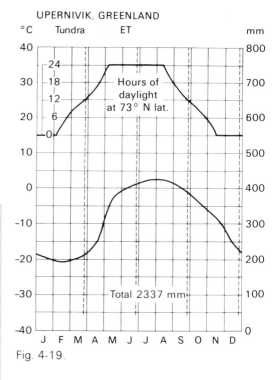

UPERNIVIK, GREENLAND

Fig. 4-19.

HIGHLAND CLIMATES

4 Other factors besides temperature and precipitation become important in elevated areas, particularly those over 3000 m. What are they and what effect do they have?

5 Because the air becomes thinner with altitude, the intensity of solar radiation increases. At the same time the temperature of the air decreases by approximately 6.4°C/km. Summarize the explanations for this apparent paradox.

6 Why is the rapid vertical change of temperature in highland areas of greater importance to man in the tropics than in areas farther polewards?

7 As altitude increases, the mean monthly temperature generally decreases and the daily range becomes greater. Explain.

STUDY 4-7

1 Determine the Köppen climate group to which each station in Fig. 4-20 belongs.

STATION			J	F	M	A	M	J	J	A	S	O	N	D	Total
1	(170 m)	T	0	2	7	13	19	24	26	25	22	15	8	2	
		P	58	63	88	95	113	113	88	85	80	73	70	63	989
2	(2 m)	T	26	26	27	27	27	27	27	28	28	28	26	26	
		P	200	113	172	137	285	297	250	172	80	75	152	283	2216
3	(160 m)	T	−45	−36	−23	−9	4	14	17	14	6	−8	−28	−41	
		P	8	5	3	8	10	28	40	33	25	13	10	8	191
4	(33 m)	T	10	12	15	22	28	32	34	34	31	25	18	12	
		P	23	25	28	13	3	<3	<3	<3	<3	<3	20	25	152
5	(196 m)	T	27	28	28	28	27	26	24	24	25	26	27	27	
		P	8	23	75	123	143	160	130	73	168	150	43	10	1106
6	(7 m)	T	−26	−28	−26	−18	−7	1	4	4	−1	−8	−18	−23	
		P	5	3	3	3	3	8	23	20	13	13	8	5	107
7	(512 m)	T	21	20	18	16	11	9	9	10	12	15	17	19	
		P	3	3	5	13	63	83	75	55	30	15	8	5	358
8	(5 m)	T	25	26	28	30	29	27	27	27	27	28	27	25	
		P	3	5	8	50	303	473	572	520	388	178	68	10	2578
9	(848 m)	T	0	1	5	11	16	18	22	23	18	13	8	2	
		P	33	30	33	33	48	25	13	10	18	23	33	48	347
10	(263 m)	T	−19	−17	−9	3	11	17	19	17	12	5	−5	−14	
		P	23	23	30	35	58	78	78	63	58	38	28	23	535

Fig. 4-20.

2 The stations in Fig. 4-20 are relatively large cities that are shown in most atlases. With the help of Figs. 3-25, 3-26, 3-27, and Fig. 4-1 attempt to identify each of them. (The name of each place is found at the end of this chapter.)

3 Prepare an analysis of the factors influencing the particular climatic characteristics of each place.

CLIMATIC CHANGE

Day-to-day variations in the weather affect most parts of the earth. However, when we refer to the climate of an area we think of something fairly stable, as the previous classification would suggest. It is true that in any area one winter may be colder than another, or one summer particularly wet, but these differences tend to average out. Such abnor-

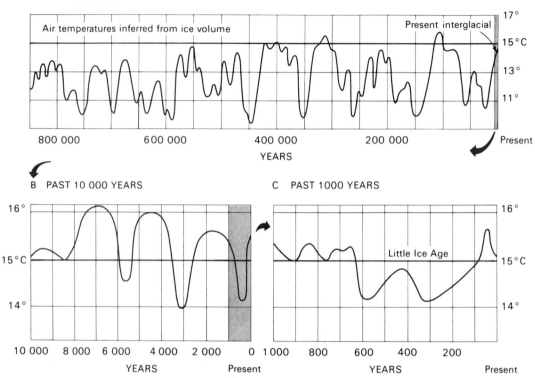

A PAST 800 000 YEARS

B PAST 10 000 YEARS   C PAST 1000 YEARS

Fig. 4-21. Average atmospheric temperatures at the earth's surface estimated for the past 800 000 years. Note: the three graphs are drawn at different scales.

malities in the weather are to be expected; they are a part of the 'normal' climate. On the other hand, we have known for some time that the earth has experienced periods of major glaciation during which the climates of the earth were considerably different than they are today. Recently it has been discovered that climatic change has taken place on a much smaller time scale than the one associated with major periods of glaciation. The discovery that such short-run changes may have serious consequences on the human population has generated widespread interest in the subject of climatic change. Scientists from a variety of disciplines are now studying the nature of climatic change. The first results of their studies have confirmed that the earth's climate does change, and often quite

rapidly (see Fig. 4-21). What they have not yet discovered is the long-term direction in which the climate of the earth is presently changing and the specific combination of factors that are responsible for these developments.

It is believed that the nature of past climates has had a great bearing on human history and development. Mass migrations of people as well as the decline and fall of civilizations over the past hundreds and thousands of years appear to have some links with changing climates. For example, the invention of agriculture and later the rise of the first great civilizations in Egypt, Sumer, and India occurred during a warm period that began about 8000 years ago and lasted for almost 3000 years. Much drier conditions followed and lasted for many centuries, contributing to the destruction of some of these civilizations while dispersing and reducing population in others. However, when these adverse climatic events occurred in times past, the density of people on the earth's surface was very small compared with the present. There was room for people to manoeuvre, thus escaping the harmful changes in climate that affected certain regions. With our current world population in excess of four billion, we have lost this flexibility. For example, during 1972 a number of weather-related disasters affected food production in many parts of the world. While the industrialized countries were able to compensate for any food deficiencies by using their food reserves, substituting one food for another or trading with countries not affected, the developing countries did not always have the same options. Consequently, in parts of Africa and southern Asia millions of people suffered and many died from starvation.

What is happening to the world's climate today? Some scientists believe that the earth is moving into a period of lower temperatures. Some have even suggested that we are at the beginning of a new ice age. While the actual amount of cooling has been small, about 0.5°C since 1940, such a change may be more serious than it appears. The most disturbing discovery is that extreme and unpredictable weather events—such as droughts, floods, and temperature extremes —are more probable during cooler periods.

Past Climatic Changes

The record of past climatic changes currently being compiled is far from complete. However there is enough evidence now available for a general reconstruction of past temperature conditions, as shown in Fig. 4-21.

Many methods have been used to establish the temperature data illustrated in Fig. 4-21. For example, fossil pollen in ancient peat beds has provided an important clue to past climatic conditions by indicating the type of plants that grew in a specific locality. Tree rings have also been used as an indicator: bristlecone pines in the western United States, the oldest living organisms in the world, have provided information about weather conditions thousands of years ago. Core samples taken from ice layers deep in ice caps, such as those covering Greenland and Antarctica, indicate the nature of air temperatures dating back thousands of years. Similarly, core samples from the sediments at the bottom of the ocean reveal the types of sea creatures that once lived in the ocean. Such information, which provides a good indication of previous ocean temperatures, has enabled scientists to make estimates of the air temperature for periods extending back millions of years. Of course these and other methods involve very complex procedures. However, after years of scientific research, and the checking of the results of one type of research against another, common temperature patterns have begun to emerge.

While one of the ultimate objects of these studies is to enable scientists to understand the causes of climatic change and thereby predict future climates, one of the more interesting results of this research has been to show the relationship between climate and past events in human history. For example, we now know that the climate warmed up during the 9th and 10th centuries, a period during which the Vikings crossed the North Atlantic and established settlements in Iceland, Greenland, and the eastern shores of North America. Early in the 13th century temperatures were again on the downward trend, and by the 15th century the Greenland colony had disappeared and the world entered a climatic phase sometimes referred to as 'the little ice age'. Various kinds of records available from this period provide considerable evidence of the colder conditions: expanding glaciers, famine, plague, and the freezing of rivers and bodies of water that seldom froze over before.

By the middle of the 19th century a warming trend had begun. From that time to the present—a period generally considered to be one of the warmest in the past 4000 years—the world's population increased from about 900 million in 1800 to over 4 billion in 1977. To accommodate these great numbers, human settlements and agriculture have spread into areas that were unsettled and could not have been cultivated two or three centuries ago. Because of the great increase in population and the large area of land required to produce the food to support this population, it is crucial that we know where we stand now in terms of climatic change. Are we at the beginning of a cooler period or is the present cooling trend just a temporary pause in the warmer weather that began in the 1800s? Once again the answer is simply that no one knows. Scientists have not yet discovered the specific combination of factors causing climatic fluctuations, nor do they know if such changes are cyclical or erratic events. Thus at the present time an increasing amount of research is being directed towards establishing the causes of climatic change. When this knowledge becomes available, we may then be able to predict future climates and it is hoped make the necessary adjustments to prevent the kind of human tragedies experienced in 1972.

Causes of Climatic Change

There are two kinds of mechanisms responsible for climatic change. The first are mechanisms external to the climatic system. These include the following: changes in the output of solar energy; changes within the earth—for example, the effect of volcanic activity; and, particularly, changes caused by human activity. The second, internal mechanisms,

involve changes brought about by interactions within the climatic system, that is, between the atmosphere and those elements of the biosphere with which it interacts, such as the oceans, land surfaces, ice surfaces, and vegetation cover. A change in one of these elements will bring about a change in the atmosphere which in turn will affect the first element in a kind of feedback relationship. The examples that follow in the section on internal mechanisms should make this clearer. It is probably safe to say that most of these mechanisms are not completely understood. Furthermore, when the proper explanation for climatic change is finally established, it will probably reveal that most—if not all—of the factors discussed interact with each other and together are responsible for the changes in climate that the earth experiences.

The extent to which climatic variations are caused by internal and external mechanisms is not yet clear. In all likelihood internal mechanisms are responsible for changes on a time scale of years to decades, while external mechanisms are important in relation to changes on time scales of hundreds and thousands of years.

CHANGE DUE TO EXTERNAL MECHANISMS

SOLAR OUTPUT

Variations in the total output of solar energy have been suggested as one possible cause of climatic variation. We have little idea of the extent to which such variations occurred in the past, and it is difficult to measure them in the present. For example, there is a long history of attempting to relate changes in solar output to sunspot activity. This record has well-defined cycles of 11, 22, and 44 years. While certain weather changes have been recorded during these cycles (such as temperatures higher or lower than average during the time of minimum sunspot activity), a direct causal connection has not been proven.

Over much longer periods of time fluctuation in solar output is thought to be related to energy-generating forces within the sun and the passage of the sun through clouds of cosmic dust or interstellar matter. Many hypotheses exist suggesting variations in solar output at intervals of hundreds of millions of years. While most scientists would agree that there is a strong probability that such variations do occur, and that they cause climatic changes on the earth, there is insufficient direct evidence to prove any of these theories.

PARTICULATE MATTER

Particulate matter (aerosols) in the atmosphere may affect the earth's radiation balance as well as the formation of clouds and precipitation from the condensation nuclei that particles provide. The major sources of these particles include: salt particles from wind-driven spray and breaking waves, volcanic eruptions, dust stirred up by wind storms, ash from industry, smoke from fires, and the release of various gases of both natural and human origin (for example, sulphur dioxide). Such particles will prevent some solar radiation from getting through to the earth. If this occurs in the upper atmosphere the earth will get cooler, but if the particles are concentrated mainly in the mid or lower atmosphere, warmer temperatures will normally follow. The effects of major volcanic eruptions on the earth's radiation balance have been observed. For example, after the Krakatoa eruption of 1883 (see p. 184) precipitation generally increased and it was estimated that world temperatures dropped about 1 °C for a few years. However, since it is difficult to separate the effect of volcanic eruptions from other factors contributing to climatic variability, we do not know if they alone caused these changes.

CARBON DIOXIDE

Most scientists agree that there has been an increase of about 10 per cent in the amount of carbon dioxide (CO_2) in the atmosphere since the latter part of the 19th century. Since CO_2 absorbs more long-wave earth radiation than short-wave solar radiation (see p. 10), this should result in an increase in temperature in the troposphere. Whether it has or not seems open to question, particularly since world temperatures have been declining since 1940. However, a doubling of atmospheric CO_2 is considered possible by early in the next century. According to some estimates this could cause world surface temperatures to increase by as much as 1.6°C. There is some doubt about this estimate because many scientists believe that CO_2 will decline as the use of fossil fuels as a major energy source declines over the next few decades. Until then, however, the possibility that such an increase could occur merits close attention.

OZONE

Some experts have predicted that the use of fluorocarbon aerosol propellants (in items such as hairsprays), chemicals used in refrigerants, and nitrogen oxide exhaust from jets will lead to depletion of the atmospheric ozone shield (see p. 49). They contend that the depletion of this zone by possibly as much as 13 to 20 per cent by the year 2000 could cause a 20 to 40 per cent increase in the incidence of skin cancer, damage to plants (including plankton), and adverse effects on the weather. It would also lead to a slightly lower surface temperature on the earth. While there is considerable disagreement about how much we are affecting the ozone layer, most authorities would agree that even the possibility of depleting this layer justifies concern.

OTHER CAUSES

Human interference with or manipulation of

the environment can occur in other ways. Examples include: the release of great quantities of heat from cities; changes to the energy and water cycles as a result of vegetation destruction; poor land-use practices; and increased irrigation, particularly that involving the diversion of major river systems. All these factors affect local climates and quite possibly also contribute to large-scale climatic changes.

CHANGE DUE TO INTERNAL MECHANISMS

Any factor that tends to warm or cool the climate may have its warming or cooling effect amplified by a kind of feedback relationship.

WATER VAPOUR FEEDBACK

The ability of the atmosphere to hold water vapour depends on the temperature of the air. If temperatures increase, for example, as a result of an increase in CO_2, then the water vapour content of the atmosphere will increase. When this occurs, the transparency of the lower troposphere to earth radiation is reduced and temperatures rise higher. This condition can produce a further increase in water vapour, which will in turn amplify the warming effect.

ICE COVER FEEDBACK

If world surface temperatures were lowered for any reason, the amount of snow and ice cover on the earth would increase. Because of the high reflectivity (albedo) of these surfaces, the amount of solar energy absorbed by the earth-atmosphere system would decrease, thereby causing a further lowering of temperature.

SEA-SURFACE TEMPERATURE VARIATIONS

Sea-surface temperatures in any locality may sometimes deviate significantly from long-term averages for periods of weeks or even years. These variations are among the most frequently mentioned factors affecting the behaviour of the atmosphere. While not fully understood, there appears to be a feedback relationship between the atmosphere and the oceans whereby atmospheric temperature variations and sea-surface temperature variations reinforce one another. Since many of the major air masses affecting continents originate over oceans, changes in sea-surface temperatures can cause significant climatic fluctuations over land surfaces.

LAND SURFACE FEEDBACK

Feedback between the atmosphere and the land surface and its plant cover is often important in times of drought (see p. 30). For example, a fairly large and lengthy decline in precipitation will result in the deterioration of the plant cover. Dust stirred up during such a dry period may cause the clouds to become over-seeded, that is, create many small droplets that cannot readily come together to form drops large enough to produce rain. This condition tends to reinforce the drought. Only a long period of rainfall caused by forces external to the drought area can bring this type of weather to an end.

This list of external and internal mechanisms is far from complete. Other factors that should be noted include: variations in the earth's orbit around the sun, changes in the angle formed between the axis and the plane of orbit (thought to vary between 21.5° and 24.5°), the effect of continental movement (see section on Continental Drift, p. 168), and changes in the position of the polar axis.

It is important to emphasize that investigation of the causes of climatic change is extremely complex, as the foregoing summary suggests. In the past, natural causes alone were responsible for climatic change. However, analysing these factors has been greatly complicated by the impact of human development on the atmosphere, particularly in recent decades.

1 When plotted on the same scale as graphs B and C in Fig. 4-21, the following temperature figures (°C) will give a general picture of the changes in world surface temperatures over the past 90 years:

1880	14.7	1890	14.8	1900	15.1
1910	14.9	1920	15.2	1930	15.3
1940	15.4	1950	15.2	1960	15.2
1970	15.1	1977	15.0		

From Fig. 4-21 and the statistics above, describe the general temperature levels over the past few decades compared with: the past several hundred years, the past several thousand years, and the past several hundred thousand years.

2 One of the principal aims of climatic research is to predict climatic change in the future. Why then has so much effort gone into the restructuring of past climates? Climate change may be either cyclical or erratic. Using the materials in this section, state which of these explanations you prefer and why. Suggest some of the reasons why it is difficult to establish the causes of climatic change.

3 For the purposes of this question assume that the earth's surface temperatures are falling. Discuss the implications of this in view of each of the following situations or statements.

a The world's population has doubled in the past few decades and if present trends continue will double again early in the 21st century.

b The world stockpiles of basic foods, such as wheat and rice, have decreased considerably in the past few years. Recent studies on the American Midwest have suggested that high crop yields may be more a result of good weather conditions over the past 15 years or so than of scientific and technological advances in agriculture.

c 'The real crisis is not so much a climate crisis, but a lack of world leadership capable of devising and agreeing on a world food security plan.' (Statement by S. Schneider of the U.S. National Centre for Atmospheric Research.)

AIR POLLUTION

It is appropriate to end this section by examining the problem of air pollution. Simply stated, this is the manner in which man and nature are introducing harmful substances into the atmosphere and thus altering its composition. As we have seen in the previous section, such changes in the atmosphere are considered to be significant factors in climatic change. In addition, it has been shown that air pollutants have a serious impact on the well-being of all living matter in the biosphere.

Air is an essential resource for sustaining life on earth. It is a renewable and seemingly inexhaustible substance available in equal amounts on all parts of the earth's surface. In addition to its primary function of supporting life, air has many secondary functions. These include the use of air as a transportation and communication medium. For example, it is an important component of the hydrologic cycle (see p. 21) that conveys water from the oceans to the continents. Air is a necessary part of many manufacturing processes and also serves as a receptacle for the disposal of waste materials from human activity. Until recently it was believed that the restorative ability of the atmosphere was limitless, and that we could dump whatever we wanted in the air and it would be dispersed. However, the growth of population and industry in recent decades—and particularly their concentration in large urban centres—has proved otherwise. We now know that the capability of the atmosphere to handle waste materials can be exceeded. Such excesses interfere with the air's primary function, that is, to support life.

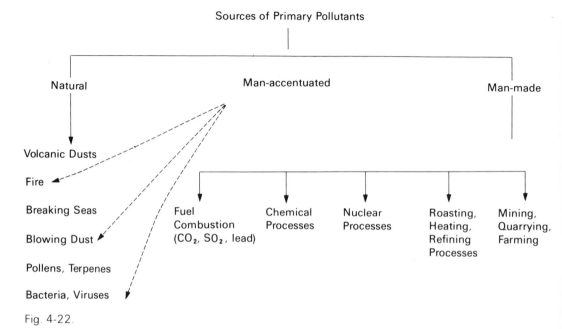

Fig. 4-22.

Production of Pollutants

Pollutants can be formed in two different ways. *Primary pollutants* are those emitted from identifiable sources on the ground—for example, chimneys, car exhausts, or volcanoes. *Secondary pollutants* are new substances formed in the air by the interaction between primary pollutants and the normal components of the atmosphere.

The primary pollutants shown in Fig. 4-22 may occur either as gases or in a particulate state. Although the discharge of particulate matter may have many harmful effects, these particles may also serve as nuclei around which water vapour condenses, producing fog, or haze conditions. It has been proven, for example, that cloud cover, fog (particularly in winter), and precipitation are greater over urban areas than adjoining rural areas because of the greater emission of pollutants over cities. Gaseous pollutants include carbon monoxide (CO), sulphur dioxides (SO_2), oxides of nitrogen (NO, NO_2, NO_3), and hydrocarbons. (Carbon dioxide falls in this

category but is not usually regarded as a pollutant.) While such substances themselves cause damage they are also the basis for secondary forms of pollution. One type of secondary pollution occurs when significant amounts of sulphur dioxide are converted by the atmosphere to sulphur trioxide, which in turn combines with water vapour to form droplets of sulphuric acid. When these droplets are covered with soot from oil or coal, they produce a damaging 'pea-soup' smog. An infamous example of such a fog occurred in London, England, from 5 to 8 December 1952. It began when a temperature inversion developed over the city. The stagnant air became increasingly polluted as automobiles, industries, electrical generating plants, and home furnaces and fireplaces continued to pump their wastes into the air. The principal pollutants were particulate smoke, sulphur dioxide, and carbon dioxide, most of which were produced by burning coal. Visibility was greatly reduced—in some theatres the screen could be seen only from the first few rows. Of greater consequence was the suffering of people who had bronchitis, broncho-pneumonia, and heart disease, conditions made worse by the pollution. Deaths during the week were 4000 above normal. Because of restrictions later imposed on burning coal, it is thought unlikely that such a severe fog could occur again.

Another type of secondary pollution results from the reaction of various nitrogen oxides and hydrocarbons (from automobile exhausts) with sunlight, forming ozone and other oxidation products. Like other pollutants these cause haze, eye irritation, and respiratory problems, as well as damage to plants by causing a reduction in photosynthesis. This type of air pollution, properly referred to as a photochemical smog, is common to most large cities. The city of Los Angeles is particularly affected because of the number of automobiles, the frequency of temperature inversions, and the abundance of sunshine in this area. Despite rigid pollution controls there has been little improvement of the smog condition in Los Angeles since 1960. As the population of the city has grown the total volume of pollution has increased, though the amount of pollution per person has declined.

Concentration of Pollutants

The rate at which gaseous and particulate material is dispersed by the atmosphere depends on both the vertical (turbulence) and horizontal (wind) motion of air. While variations in the latter affect the pollution content of the air at any given time, the former is often the chief factor responsible for serious pollution problems.

The type of turbulence referred to as convective turbulence depends on the vertical change in temperature (lapse rate) in the atmosphere at any given time. If the air at the surface of the earth is warmer—and therefore less dense—than the surrounding air, it will tend to rise. The greater the lapse rate, the higher and more rapidly the air will rise,

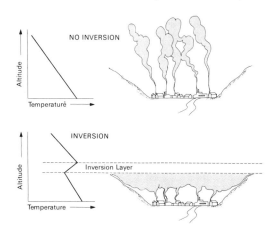

Fig. 4-23. The effect of a temperature inversion trapping pollutants.

carrying with it various pollutants. Under such conditions the atmosphere is described as unstable (p. 59). If, on the other hand, the lapse rate is small, surface air will tend to resist upward movement. The atmosphere is then described as stable. In certain circumstances the temperature of air increases with altitude for at least a short distance and a condition of strong stability (temperature inversion) is created (see Fig. 4-23). Under stable conditions the dispersion of pollutants is hindered, and when a temperature inversion persists for several days pollution may become particularly serious.

The stability of air over any particular location can change rapidly. Often the change occurs between day and night. During the day the heat of the earth's surface contributes to instability; at night the cooling of the earth's surface through radiation will cause the lower layers of air to cool rapidly and produce a stable situation. Consequently, the accumulation of pollutants over a particular location may be greater during the night. This may be apparent in an early morning haze that gradually disperses as the sun heats the earth's surface.

Vertical and horizontal air movements are affected by many complex factors. These include the influence of moving high (anticyclones) and low (cyclones) pressure systems, particularly in the middle latitudes. Low pressure systems are generally associated with high wind speeds and ascending air, which help disperse pollutants. High pressure systems, on the other hand, are normally slow moving and are often accompanied by lower wind speeds. In addition the slow sinking, and consequently warming, of air produces stable conditions. If these conditions persist for several days, as they may, pollutants become concentrated. Also affecting the concentration of pollutants is the terrain of a particular area. For example, river valleys or basins surrounded by hills or mountains tend to experience the gravitational flow of colder air into the valley or basin. In addition, average wind speeds in such areas may be lower because of their sheltered location.

We have tended to regard the effects of air pollution as a relatively localized phenomenon. Indeed, its effects are greatest in and around our large urban-industrial centres. In a recent study,[4] however, precipitation samples were collected in many locations in southern Ontario, some of which were hundreds of kilometres from major cities. When these samples were tested for various pollutants they showed few significant differences, that is, all of them were more or less equally polluted. The pollutants discovered—and their quantity in grams per hectare per year —include: sulphate (42 000), nitrogen (21 000), sulphite (16 000), potassium (7700), calcium (7000), chloride (5400), sodium (4000), phosphate (3000), magnesium (1600), zinc (565), lead (58), copper (58), and PCBS (0.7). From this and other studies it is becoming increasingly apparent that pollution, whether in the air or the water, cannot be localized. No one is immune from its effects, and thus everyone should be concerned with its prevention.

STUDY 4-9

Discuss the problems created as well as the difficulties involved in solving the following four aspects of air pollution.

1 Pollution tends to increase in quantity and complexity as production and consumption increase, that is, as a society becomes increasingly industrialized and living standards rise.

2 While air pollution is concentrated in urban areas, it is dispersed by the atmosphere without regard for either the boundaries between city or country or the political boundaries separating states.

3 Air is a public good that is free for all to use. Its primary purpose—to support life—conflicts with its use as a sink for the disposal of waste materials from human activity.

4 Air pollution control is largely a question of economics. Maintaining the quality of the atmosphere in a condition satisfactory for the life forms of the biosphere is a growing economic burden for all segments of society.

Stations in Fig. 4-20

Station	1	St. Louis, Missouri
Station	2	Georgetown, Guyana
Station	3	Yakutsk, R.S.F.S.R., U.S.S.R.
Station	4	Baghdad, Iraq
Station	5	Ibadan, Nigeria
Station	6	Barrow Point, Alaska
Station	7	Santiago, Chile
Station	8	Rangoon, Burma
Station	9	Ankara, Turkey
Station	10	Winnipeg, Manitoba

[1] H. Shirmer, Le Sahara (Paris, 1893).

[2] W.G. Kendrew, The Climates of the Continents (5th ed.; Oxford: The Clarendon Press, 1961), pp. 411-12.

[3] Ibid., p. 572.

[4] M. Sanderson, Agricultural Watershed Studies, Great Lakes Drainage Basin, Canada. Task Force C, Project 6, Precipitation Quantity and Quality. (Industrial Research Institute, University of Windsor, 1977.)

5/ VEGETATION

The soils of the earth provide nourishment for an amazing variety of plants. Some grow virtually as nature created them, while others have been cultivated and changed by humans. The latter come under the domain of agriculture, while the former, considered as natural vegetation, are dealt with here.

'Natural vegetation' can be a confusing term. All vegetation on the earth existed in a natural state until we began to use or cultivate it for various purposes. Entire regions are now covered with cultivated plants. In other regions the natural vegetation has been modified by the introduction of various plants (certain trees, grasses, or weeds) that were not cultivated. Natural vegetation can also change without human influence.

Over large parts of the earth, vegetation often remains substantially unchanged in its general characteristics for a long period of time. This happens when vegetation reaches a stage of development where it has adapted to the climate of its area, to the nature of the soil, and to the other factors that determine the conditions in which it has to live. At this stage the vegetation, known as a *plant community*, is said to have reached its *climax*. In any area the natural vegetation may be at an earlier stage of development, possibly owing to climatic change, the extrusion of molten rock, fire, or the removal of forest vegetation for commercial purposes. All of these factors, occurring at different times in the past, may account for the variety of stages that the natural vegetation in different areas goes through in its development towards climax.

STUDY 5-1

Using the following five quotations as a basis, comment on the importance of natural vegetation as an element of geography.

Plants constitute for mankind the main inexhaustible source of fuel and industrial supplies and are of fundamental importance in many different branches of industry: in drug production, brewing, pulp and paper making; in lumbering, in the textile industries, in tanning, dyeing and curing; in the production of plastics, animal feedstuffs, scent, oil, rubber, resin, gum, wax, and fibres; and, of course, in the wider fields of agriculture, horticulture, forestry, fish-culture, and the direct uses of their innumerable products.[1]

As a resource in another form, it is the enduring attraction of woods, forest, and wildlife, which gives many of the most popular summer and winter playgrounds their special lure for tourists.[2]

But the forest is more than a warehouse for man's material needs. Its protective covering is renowned as a conservator of soil and as a moderator of local climate.[3]

Ranking high in importance is the esthetic contribution which it makes to the variety in appearance and the attractiveness of the earth's land surfaces, for the visual landscape is to a significant degree the product of the vegetation mantle.[4]

The significance of vegetation to geographers . . . rests above all on its far-reaching indicator value . . . a plant community with its characteristic floristic composition and its particular set of life forms is not only a sure indicator of the coarser environment and historical aspects of the landscapes, but it unerringly points out the more subtle and hidden characteristics as well.[5]

THE ELEMENTS AND CLASSIFICATION OF VEGETATION

In studying vegetation, the geographer is concerned principally with its overall character rather than with the sort of detailed classification that a botanist uses to identify each individual species. There are four major classes of natural vegetation that occur on the land areas of the earth: forest, grassland, desert, and scrub. These four classes or plant communities can in turn be divided into many different sub-communities. While the major plant communities are not difficult to recognize and distinguish from each other, a more comprehensive breakdown into various sub-communities presents difficulties similar to those experienced in trying to classify most geographic elements: the detailed pattern is very complicated. Vegetation consists of a great number of individual plants, and the combinations they can form are infinite.

Listed below are the main aspects of vegetation that help distinguish the various communities and sub-communities. This list will serve as a basis for observing some of the more important characteristics of the different types of natural vegetation.

1 Major species of plants.
2 Characteristics of the leaves, stems, and roots of the major species.
3 The overall appearance of the vegetation including its arrangement, density, height, vertical stratification, and colour.

Using these criteria, it is possible to subdivide the four main communities of plants into innumerable sub-communities. The major ones are shown below and are sufficient for a general description of the pattern of the world's natural vegetation.

1 FOREST COMMUNITY
   Tropical rain forest
   Lighter tropical forest
   Broadleaf and mixed broadleaf-coniferous forest
   Coniferous or Needle-leaf forest
2 GRASSLAND COMMUNITY
   Tropical savanna
   Middle-latitude grasslands (prairie and steppe)
3 DESERT COMMUNITY
   Desert scrub and barrens
   Tundra
4 SCRUB COMMUNITY
   Tropical scrubland
   Mediterranean woodland and scrub

THE CONTROLS OF VEGETATION

The physical environment causes variations in the patterns of distribution of the main types of natural vegetation. This environment is made up of interacting elements such as solar energy, climate, bedrock, soils, drainage, landforms, and wild fauna—all of which are parts of the ecosystem. Unlike animals, plants are not mobile and cannot escape their environment. They must adapt to the conditions that an area imposes upon them. Some of the most important of these conditions are described below.

Heat

Every plant has a maximum and a minimum temperature beyond which it cannot live and an optimum temperature at which growth occurs more quickly. The maximum is of least importance because no area on earth is so hot that a great variety of plants cannot survive given a sufficient amount of moisture. The minimum limit, however, is far more restrictive. Substantial reduction in species of plants occurs as temperatures and the length of the growing season decrease towards the poles, or as elevation increases. For example, the 10°C isotherm for the warmest month (Fig. 3-26) is the approximate location of the *tree line*, beyond which the growing season is too short and too cool for trees to develop. Here a tundra or alpine type of vegetation begins. Plants adapt to areas with marked cold seasons by reacting to the cold in one of the following ways:

1 Deciduous plants stop growing and shed their leaves.
2 Evergreen plants stop growing but undergo no visible change during the cold season.
3 Other plants (known as *annuals*) die, and the species is perpetuated by seeds that can survive the cold. Some grasses and many plants referred to a weeds are examples of annuals.

Water

All plants require water. It is the principal ingredient of sap, the watery solution of minerals and organic products that feeds all parts of the plant. In the process of sustaining itself a plant takes in and loses large amounts of water, especially through its leaves. This process is called *transpiration*.

Both the species of plants and their charac-teristics vary according to the amount of water available. Plants that exist in water or in very damp and humid areas are known as *hygrophytes* (from the Greek *hygro*, meaning wet). Normally they have shallow roots, large thin leaves, and long thin stems. The water lily is an example of a hygrophyte. Plants that survive in very dry areas are known as *xerophytes* (from the Greek *xero*, meaning dry). They share common characteristics that result from having to withstand severe drought conditions. These characteristics include: deep and widespread root systems; short thick stems and small thick leaves; and a thick bark or coating of wax on the leaves and stems, which prevents the loss of water. The cactus is an example of a xerophyte.

In climates that have a wet and a dry period, or in climates with a distinct cold period, many plants are hygrophytic at one time and xerophytic at another. These are referred to as *tropophytes* (from the Greek *tropos*, meaning change). For example, in tropical areas with a distinct dry season, most trees discard their leaves during the period of drought, their woody stems and branches conserving water; when the rains come, buds open, and new leaves are formed. In the middle latitudes, most deciduous trees, such as the oak or maple, are tropophytes.

Soil

Soil has an important influence on vegetation. However this influence is complicated by the fact that the soil is itself a result of an interaction of temperature, precipitation, and vegetation on the surface rocks of the earth's crust. While climate is largely responsible for establishing the major types of vegetation, the innumerable local variations within any type are to a great extent determined by soil.

Local variations in soil are caused by differences in the depth and nature of the weathered bedrock from which the soil is formed, and differences in drainage, slope, and type of bedrock. These differences in turn cause local variations in vegetation. For example, sandy or stony soils—which are very porous —may produce a xerophytic vegetation, even in regions of moderate rainfall.

Competition

In spite of the fact that the heat, mosisture, and soil conditions in most localities allow for a great variety of plant species, less variety exists than one might expect. The reason for this is competition between plants. Through the evolutionary process of natural selection, certain plants become dominant and eventually establish a climax vegetation. A simple example of competition can be seen in the relationship between trees and grasses. In many parts of the world the existence of a naturally growing forest has precluded the growth of grasses, most of which cannot survive in the shade of trees. However when trees are cut down in such areas—for example as in England or New Zealand—the land has proved to be suitable for the growth of pasture grasses.

Competition occurs when there is not enough of some resource for two individuals or two species in a community. The concept of niche differences may be used to help us understand the importance of competition in explaining the distribution of plant species and communities. A niche is the place of one species in a plant community in relation to other species. The idea is connected closely to the principle of specialization, which is also relevant in many branches of human geography. By specializing in their jobs, people make best possible use of their inherent and developed capabilities and avoid some competition. Specialization in plant communities occurs for similar reasons. Thus individual plants in the same community complement one another in different ways: by drawing on different resources, by using different segments of space in the environment, and by

utilizing the same resources at different times of the year.

Competition for light in forest communities illustrates this principle. Tall trees with large canopies reduce the intensity of light for plants growing at lower levels. As a result the growth of young trees that germinate in this shade is limited, and many die. However smaller species, such as shrubs or flowering plants, can successfully occupy these shady areas. In a deciduous forest, for example, smaller species either adapt to living in conditions of low light intensity or have their growth period before the trees develop foliage.

Competition and niche differences are also evident during the evolutionary stages in the development of a plant community. For example, when an area of middle-latitude broadleaf forest is cleared and then allowed to regenerate, it will go through a series of stages before developing into a mature forest. During each stage different plants will be dominant. At first the cleared ground is covered by grasses and other low-growing plants. Then several species of shrubs may appear, growing as high as six metres and shading out the grasses. These shrubs provide the nursery for the dominant forest trees that germinate in their shade. Soon, however, these young trees push their way through the low canopy. Once the trees emerge, and begin to shade the lower levels, the shrubs must compete with them for light, moisture, and nutrients. In competition the shrubs become tall and spindly, more vulnerable to hazards, and some die. The shrubs become less dense and the forest that develops is a potentially permanent or climax community. Similar stages can be recognized for all plant communities.

There are several important generalizations to remember about niche differences in relation to plant communities. When two species occupy the same niche in a community, one will become extinct. It follows that no two species in a community are in direct competition. Instead, species tend to complement one another by using different aspects of a community's resources. Observing the characteristics of different plants in a community may help us to understand the functions served by each.

Human Influence

It is essential to emphasize once again the influence people have had on vegetation. We have entirely destroyed the natural vegetation in many places—especially the major agricultural areas of the world. We have also brought about other less obvious changes. For example, a large number of the smaller plants in the scrub forest of some semi-arid parts of California, which appear to be indigenous, were in fact introduced by the Spaniards during the early periods of settlement. Similarly, many of the grasses on the prairies of Canada and the United States belong to European species that were brought over by European settlers, just as thousands of years earlier these same species were introduced to Europe by neolithic immigrants from Asia.

TYPES OF NATURAL VEGETATION

Figure 5-1 shows the climax vegetation for the entire land area of the earth. However, it does not indicate those areas where climax vegetation has been largely removed or modified by people. In addition, the map is highly simplified in that it ignores all the important variations that can be found within the main types of natural vegetation. When using this map it should be kept in mind that each vegetation type consists of a patchwork of different plant communities. The boundaries between types represent areas (known as *ecotones*) where the species of one type intermingle with those of other types.

The most important division in vegetation is the division between forest and non-forest, but the boundary between the two is sometimes indistinct. The term 'forest' should therefore be restricted to areas where the trees form a more or less continuous canopy.

Forests

Approximately one-third of the world's land surface consists of ecosystems made up of forest vegetation. Trees are among the oldest and the largest living organisms. For example, some live bristlecone pines found in California are over 4000 years old, and the California redwoods reach heights of well over 100 m.

Forests can be subdivided according to several traits. The most comprehensive subdivision is probably by the characteristics of leaves, since trees with similar leaf characteristics usually have many other features in common. Tree leaves are generally either needle-like (e.g. the pine) or broad in shape (e.g. the maple). Most trees growing in the high and middle latitudes are needleleaf and, because they also bear cones, are called conifers. Trees growing naturally in middle and low latitudes are predominantly broadleaf.

Trees can be further subdivided according to their leaf-shedding habits. Evergreens are covered with foliage at all times of the year, whereas deciduous trees shed all their leaves at least once a year. Most broadleaf trees in middle latitudes are deciduous. In low latitudes, broadleaf trees—such as the laurel or locust—are evergreen. Almost all coniferous or needleleaf trees are evergreen, with the exception of the Eurasian larch (called tamarack in North America), which is needleleaf deciduous.

TROPICAL RAINFOREST

Together with the coniferous (boreal) forest, the rainforest is one of the largest surviving forests in the world. It presently occupies about 10 per cent of the world's land area,

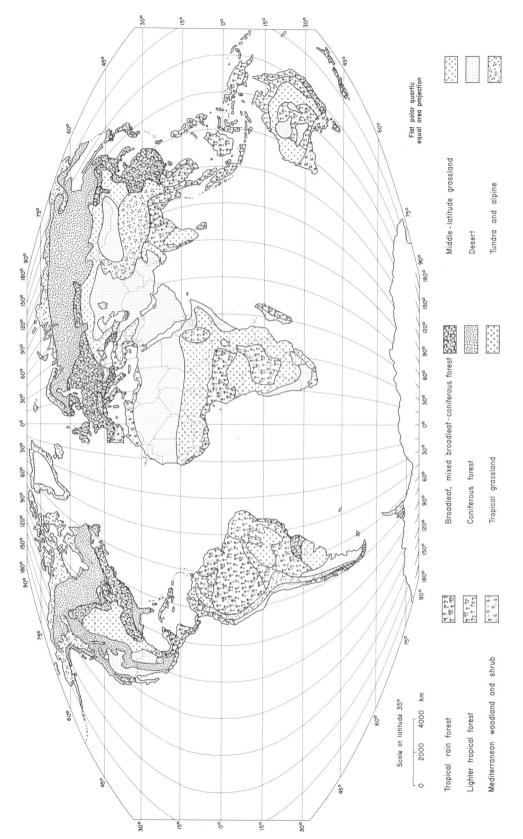

Scale at latitude 35°

0   2000   4000 km

Tropical rain forest

Lighter tropical forest

Mediterranean woodland and shrub

Broadleaf, mixed broadleaf-coniferous forest

Coniferous forest

Tropical grassland

Middle-latitude grassland

Desert

Tundra and alpine

Flat polar quartic
equal area projection

Fig. 5-1. Types of natural vegetation.

but this total is slowly diminishing as a result of the development of forestry and agriculture in these areas. This is particularly true of the Orinoco and Amazon basins, the two major areas of rainforest in the Americas.

The rainforest is very dense, with a tightly closed canopy. Three storeys may be distinguished. The upper storey is discontinuous, composed of a relatively few gigantic and usually isolated trees with mighty crowns rising 40 to 45 metres above the ground. At a height of about 25 to 30 metres a continuous middle storey of crowns pressed one against another gives the forest, as seen from an airplane, a characteristically undulating and unbroken appearance and hides the trunks from view. The lowest storey is made up of small trees and bushes whose crowns fill almost all of the remaining spaces. . . . The ground, however, is bare, or garnished with a few sporadic herbaceous plants. It is easy enough to walk through the forest, but creepers frequently hinder rapid progress. . . . The most remarkable characteristic of many of the great trees of these forests is the buttress structure at their bases. Usually triangular, these buttresses serve as firm anchors for very tall trees that are otherwise attached to the soil by shallow root systems only. . . . In the true rain forest the foliage is always green and never leafless. Even so, there is a vegetative rhythm in harmony with the climatic rhythm; there are periods when new leaves appear and periods when old leaves fall. . . . The rain forest is heterogeneous, composed of a complex mixture of species. One hectare may contain 50 to 90 species of trees and large shrubs, exclusive of creepers and herbaceous plants. The composition varies greatly from place to place. A forest of some thousands of hectares may include 200 or 300 species of trees and large shrubs, and a complete inventory of a whole territory would add up to several hundred more.[6]

Fig. 5-2. A sapucema tree on the edge of the Amazon River near Macapá, showing typical buttress-trunk structure.

STUDY 5-2

1 Describe briefly the environmental characteristics responsible for the tropical rainforest. Why is the area covered by rainforest smaller in Africa than it is in South America?

2 Many misconceptions about the density of the lower part of the tropical rainforest have arisen because exploration of these areas was often carried out along rivers where dense, jungle-like conditions exist. With the assistance of Fig. 5-4, explain the general lack of dense undergrowth in areas of the rainforest away from rivers.

3 Lianas, epiphytes, parasites, and saprophytes are some of the forms of plant life commonly found in rainforests. Using an outside source write a short note on the characteristics of each.

4 Although rainforests make up nearly 50 per cent of the world's forest areas, their

Fig. 5-3. An oblique aerial view of the tropical rain forest near Singapore. Note the uneven appearance of the forest, with a few giant trees standing above the rest. The variations in tone indicate different species.

commercial potential is largely undeveloped. What are some of the reasons for this?

5 Suggest reasons for the diversity of tree species in tropical rainforests (as many as 2000 in a few square kilometres). Broadleaf deciduous forests of the middle latitudes usually have fewer than ten species of tree per hectare.

6 Using the list of elements on p. 120 and the quotations, photographs, and answers to the questions above, write a description of the characteristics of the tropical rainforest.

MEDITERRANEAN WOODLAND AND SHRUB

This type of vegetation is characterized by open woodlands consisting mainly of broadleaf evergreens such as oak and eucalyptus. Many of the trees are low, stunted, and interspersed with shrubs and bushes. In areas where the forest has been removed either by humans or by fire, this shrub vegetation becomes dominant. It forms a type of vegetation known as *maquis* in Europe and *chaparral* in California. Maquis vegetation consists of a great variety of species including

Fig. 5-4. Light intensity in a tropical rainforest.

Fig. 5-5. Characteristic chaparral vegetation in the Los Padres National Forest, California.

olive, myrtle, lavender, and rosemary. Although seldom more than three metres high, these plants often form dense and tangled thickets. (Areas covered with maquis provided excellent hiding places for members of the French Resistance during the Second World War, and the term 'Maquis' came to be applied to these people.) In the driest Mediterranean areas—particularly those underlain by limestone rock—the original woodland has been replaced by even lower-growing plants, including prickly dwarf shrubs and scrub oak. This type of vegetation is called *garrigue*, a Provençal term for low, stunted evergreen shrubs.

In order to survive under difficult climate conditions, plants in Mediterranean areas have developed some unusual characteristics: small, thick, leathery leaves; gnarled trunks usually encased in a thick bark; and deep and wide-spread root systems. This type of root system permits the plant to propogate itself by means of suckers. Thus, even when the tops of the trees are destroyed by fire, which is a common occurrence in the dry Mediterranean summers, the roots continue to live, and a dense thicket of brush will soon emerge.

STUDY 5-3

1 Examine the distribution of Mediterranean woodland and shrub vegetation and describe the climatic condition under which it has developed.
2 Why is there so much variation from place to place in the vegetation of Mediterranean areas?
3 Using the list of elements on p. 120, describe the type of vegetation illustrated in Fig. 5-5.

BROADLEAF AND MIXED BROADLEAF-CONIFEROUS FOREST

Summer-green forests, dominated by broad-leafed trees that lose their leaves during winter, make up the main climax formation over much of temperate Europe, eastern Asia, and North America. They also reappear in some comparable regions of the southern hemisphere. From the physiological point of view the cold winter tends to be a dry period, owing to the fact that low temperatures often hinder absorption of water by the roots: this is counterbalanced by the leafless condition during winter, for it is chiefly from the leaves that loss of water takes place, and it is mainly such loss which has to be made good by absorption from the soil. If active transpiration continued when the resultant water-deficiency could not be made good by absorption, owing for example to warm weather when the soil remained frozen, serious injury and even death might result. As it is, these deciduous broadleaf forests occupy many of the most populous regions of the world, and although in such areas we may now see around us only patches of anything approaching a climax, this situation is largely due to human disturbance—to clearance for agricultural or other purposes, or to the depredations of Man's domestic animals.[7]

The North American mixed forest contains such evergreen species as red and white pine, spruce, and hemlock; the most important deciduous trees include yellow birch, sugar maple, oak, basswood, and hickory. As a result of human destruction, probably less than 10 per cent of the original forest remains.

STUDY 5-4

1 One of the main characteristics of this forest is the change in appearance that occurs from summer to winter. What are the reasons for the complete loss of foliage during the cold season?
2 The boundary of this forest zone, like the boundaries of all vegetation zones, is an area of transition. Referring to Fig. 5-1, describe

Fig. 5-6. Mixed broadleaf-coniferous woodlot in southern Ontario.[1]

the nature of the transitions that take place along the northern, western, and southern boundaries of this forest zone in North America.

3 Why has most of the mid-latitude broadleaf forest been cut down? Suggest reasons for the location of the remnants of broadleaf forests. What are some of the consequences of the removal of the forest?

CONIFEROUS FOREST

In North America there are several different areas of coniferous forest. In the nothern part of the continent, covering a substantial part of Canada, is the boreal forest—boreal meaning of or pertaining to the north. (This same type of forest covers a much greater area at similar latitudes in the USSR, where it is known as the *taiga*—from the Russian for forest). In western North America the coniferous forest extends down the coast from Alaska to northern California and can be subdivided into several different types including coastal, sub-alpine, and montane forests. Another important area of coniferous vegetation is the pine forests of the southeastern United States.

THE BOREAL FOREST OF NORTH AMERICA

A journey through certain portions of the virgin boreal forest can leave one with the impression of extreme monotony. For mile upon mile the dense cover of evergreen conifers casts an unbroken shadow on the ground beneath. In consequence undergrowth is scanty to non-existent, the ground being covered with a litter of needle-shaped leaves and dead, decaying wood. . . . The coniferous evergreen tree has proved itself to be particularly well adapted to those areas with cold winters and short growing seasons in Eurasia and North America. The reduced area of the needle-shaped leaves enables these trees to reduce water-loss by transpiration *to a very low level during the winter when water is unobtainable from the frozen soil. On the other hand, as soon as air and soil temperatures are sufficiently high to permit* photosynthesis *and other physiological processes to take place in the spring, these trees are able to profit by it.*[8]

The predominant trees in the east are the black spruce, jackpine, balsam fir, white spruce, and larch, in that order. Moving towards the west these species become less dominant and give way to the lodgepole pine, alpine fir, and others. Common broadleaf deciduous trees found throughout the boreal forest are the larch, which is a needleleaf deciduous tree, the white birch, and the aspen.

Figure 5-8 is a stereoscopic vertical view of a small area of boreal forest in the vicinity of Cochrane in northern Ontario. It illustrates the tendency of species to segregate themselves into pure stands unmingled with any other species. In this photograph a white line separates two common boreal species: the black spruce and the aspen. The black spruce is usually found in pure stands at lower elevations, often in areas of poor drainage. Where the stand is pure, the canopy is regular in pattern and the height is even. It can be identified (using a stereoscope) by the slender crowns with pointed tops. The aspen stand, on the other hand, has crowns with rounded tops, often with spaces between the trees.

Segregation into pure stands occurs because most of the major tree species have slightly different demands or tolerances with respect to such factors as soil, drainage, and slope. In an area where physical characteristics (for example poor drainage or steep slopes) are fairly uniform, one or two species tend to become dominant through the process of competition.

STUDY 5-5

1 Using a stereoscope identify the black

Fig. 5-7. The majority of the species in this swampy area of the Canadian Shield near Sudbury are black spruce. Some larch can be seen in the foreground and a few balsam fir in the background.

Fig. 5-8. The black-spruce stand is outlined in this stereo pair and is surrounded by aspen. The spruce are 82 years old and approximately 15 m high.

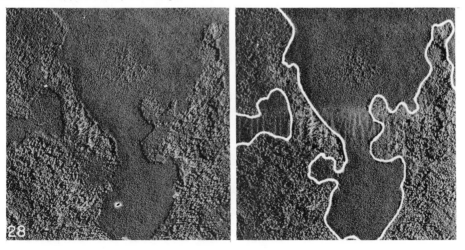

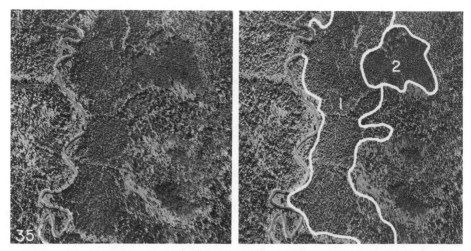

Fig. 5-9. This stereo pair was photographed in the vicinity of Kirkland Lake (Northern Ontario). The tree species are (1) balsam fir and (2) black spruce. The fir are 60 years old and 15 m high while the spruce are 98 years old and 17 m high.

spruce and aspen stands in Fig. 5-8. Describe as fully as possible the different characteristics of each as they appear on the photograph. Suggest possible explanations for the rather marked separation of the two stands. Why would the identification of the two species in this photograph be much easier in winter than in summer?

2 Describe the characteristics of the forest in Fig. 5-9. Note species, variation in species, and possible reasons for fairly large open areas. Balsam fir and black spruce stands are marked.

3 Why do needleleaf evergreen trees predominate in the higher latitudes? What prevents them from being a major species in the broadleaf deciduous forests to the south?

4 Much of the boreal forest of North America covers the Canadian Shield. What characteristics of this landform region would influence the development of the forest (Fig. 5-7)?

5 What commercial advantage results from the natural segregation of species into communities in which one or two species are dominant? In this connection, how does the boreal forest compare with the tropical rainforest? What is the principal commercial use of the boreal forest?

6 Using the list of elements on p. 120, the quotations and photographs, and the answers to the questions above, describe the characteristics of the boreal forest.

7 The area of the southern United States covered by pine forest experiences a long growing season and abundant precipitation (Fig. 5-10). What possible explanations are there for the development under these conditions of a needleleaf evergreen forest rather than a broadleaf deciduous forest? This southern pine forest is an important source of various useful materials. What advantages does it have over the forest regions to the north? How does it differ in appearance from the boreal forest?

8 The needleleaf evergreen forest in the western part of North America (Fig. 5-11) contains a variety of species. Douglas fir, Sitka spruce, western cedar, and western hemlock predominate in the northern part of this forest, and redwood, ponderosa pine,

Grasslands

TROPICAL SAVANNA

Where the length of the dry season exceeds two and a half months, the tropical rainforest gives way to two other vegetation types. In some areas it is replaced by seasonal forests and woodlands (called lighter tropical forests in Fig. 5-1) and in others by tropical savannas (see Fig. 5-12). The seasonal forests and woodlands are more open than the rainforest and also less stratified. Because they are more open, light can penetrate to the ground and grasses and shrubs are able to grow. In addition, because of the longer dry season and lower annual precipitation, trees are shorter and mainly deciduous. Tropical savannas exist in the same dry season zones as seasonal forests. They can be distinguished from seasonal forests by the fact that grasses are the predominant vegetation, although they are mixed with trees and shrubs forming open, park-like landscapes.

Authorities recognize several types of savanna. These range from the savanna woodland, where trees dominate a landscape, to true grass savannas where trees are restricted to a few scattered locations, notably along rivers. An intermediate type of savanna common over large areas of Africa, for example, is the accacia-tall grass savanna. In this savanna the main plants are tussock grasses, a metre or so in height, that form an almost continuous ground cover. The scattered trees may be deciduous, such as the accacia or combretum in Africa, or evergreen, such as the eucalyptus in Australia.

Since the seasonal forests and the savanna occur under similar climatic conditions (the Aw climate), most scientists tend to agree that there can be no such thing as a distinctive 'savanna climate'. They differ however in their explanation of the existence of the savanna vegetation types. Some claim that a combination of climatic, hydrological, and

Fig. 5-10. Eighty-year old longleaf pine in the Osceola National Forest, Florida.

white pine, and western larch predominate in the southern part. This forest is a very important source of lumber for the construction industry. Compare the western forest with the boreal forest and mention some of the reasons for its great commercial importance, particularly in coastal areas.

Fig. 5-11. Seventy-five-year old stand of Douglas fir in the Quilcene District of the state of Washington.

soil conditions would be sufficient to produce this type of vegetation. The view of the majority, however, is that frequent fires over long periods of time are responsible for destroying the trees and bushes, thus allowing grasses to take over. In Africa, for example, eye-witness accounts of great fires in these vegetation zones date back as early as the 5th century B.C. Today many of the grass savannas are burned regularly every dry season to clear away dead material and promote better grass growth in the next rainy season. In these fires the forests and undergrowth burn easily, and it is reasonable to assume that few species survive. However, grasses and a few woody plants with thick or fire-resistant bark, such as the accacia, are able to survive.

STUDY 5-6

1 Examine the distribution of seasonal forests (lighter tropical forests) and tropical savannas (Fig. 5-1) and describe the climatic conditions under which they have developed.
2 Describe some of the variations in vegetation found in savanna regions. What are some of the reasons for these variations?
3 Explain the controversy that has developed over the origin of savanna vegetation.
4 Using Fig. 5-12, discuss the relationships

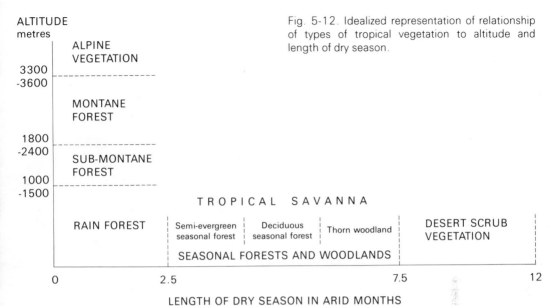

Fig. 5-12. Idealized representation of relationship of types of tropical vegetation to altitude and length of dry season.

among altitude, length of dry season, and vegetation, in the tropics.

5 In Fig. 5-12 tropical savanna and seasonal forests and woodland are represented as occupying areas with similar dry seasons. What events might lead to the replacement of forests and woodland by savanna?

MIDDLE-LATITUDE GRASSLANDS

Grasslands in the middle latitudes can be divided into two types: short grass (steppes) and long grass (prairies). The difference between the two is due largely to the amount and the effectiveness of precipitation. In most instances middle-latitude grasslands can be distinguished from their tropical counterparts by the fact that trees are found only in river valleys and along the humid margins of middle-latitude grasslands.

Grasslands occur where trees have failed to develop. While insufficient moisture is the principal explanation for the lack of trees, other factors include poor drainage, intense cold and winds, and repeated fires. Grasses are generally very hardy and can withstand such conditions, which would destroy tree growth. Much of the original vegetation in grassland areas, such as Buffalo grass and bluestems, has long since disappeared. It has been replaced with cultivated grasses, such as wheat and barley. These have given grassland regions the title of 'breadbaskets' of the world.

Deserts

The arid margins of areas with short grass gradually give way to deserts. The boundary between the two is subject to continual change. Desert vegetation varies considerably from place to place and from time to time over the course of a year, depending on how much precipitation has fallen. Some of the types of vegetation that exist in these inhospitable areas are:

1 Low grasses that form clumps, similar to those of the short-grass regions.

Fig. 5-13. Blue grama grass lightly grazed at the end of a season of exceptional rainfall (Colorado).

Fig. 5-14. Tall, dense sagebrush in Nevada. Note the spacing of the plants and the lack of any form of vegetation.

Fig. 5-15. Brush and scrubby spruce with some aspen and birch along the Yukon River in Alaska. This region represents the transition between the Boreal forest and the tundra.

2 Annual plants that germinate, grow, and produce seeds only in the few weeks following rainstorms.
3 Perennial shrubs such as sagebrush and creosote bush.
4 Water-storing plants known as *succulents*, of which the cactus family is a familiar example.

STUDY 5-7

1 What are the important changes in vegetation that can be seen over an area extending from the boundary between grassland and forest to the desert itself? What are the reasons for these changes? What are some of the characteristics of plants in the grassland and desert that enable them to grow in these regions?
2 Where in the grassland and desert regions has the natural vegetation largely disappeared? How do you account for this?
3 Why do the boundaries between tall grass, short grass, and desert vegetation types fluctuate over the years (see p. 29)? What effects does this have on the use of these areas?
4 Discuss the influence of climate and fire on the origins of middle-latitude grasslands.

The Tundra

Arctic tundra occurs in high latitudes. The various plants that make up this vegetation type flourish during the brief summers with long hours of sunlight. Since only the top few centimetres of the permafrost thaws during the summer, the flat and low-lying areas are quite marshy. *Alpine tundra* develops above the limit of tree growth and can be found on larger mountains from the equator to the pole. In its basic characteristics it is very similar to arctic tundra.

This is the land of the little plant, with only a few species growing more than ankle high. Of course, there are no upright trees; the few species north of timber line—mostly willows, elders, and birches—are stunted. Only near the treeline in sheltered valleys along flowing streams do many of the trees get head-high. But these dwarf plants often live to old age. An arctic willow may have 400 annual rings crowded into its one-inch trunk.

Apart from the short stature which they have in common, tundra plants have radically different tolerances for moisture, mineral nutrients, winter wind and cold, soil acidity and erosional forces. In the marshes, for instance, the main vegetation is likely to be a bed of soggy sphagnum moss, punctuated here and there by hummocks of a rough grass that spreads through the acid muck with underground runners. The standing marsh water forms the breeding ground for the insects that harass the tundra animals. In other wet meadows there are stands of a grasslike plant called arctic cotton, which bears its seeds in white, fluffy bolls. At the dry end of the moisture scale are vast well-drained areas covered with masses of coarse gravel and boulders. Nothing but lichens survive on the bare rock surfaces, but in the crevices small pockets of wind-blown soil and lichen remains collect and hold a little water. Moss invade these stony pockets and in time form a base on which small flowering plants can grow.

Between the extremes of marsh and gravel field are vast intermediate areas, neither drenched nor dry. Close to timber line these are often carpeted by an extraordinary yellowish lichen which is misnamed 'reindeer moss' . . . This is the prairie grass of the Arctic, furnishing winter graze for the caribou. Also blanketing many tundra hollows are the arctic varieties of some plants familiar in southern latitudes; cranberry, blueberry, and heather. In general, they can take hold only where snow accumulates in winter, giving some protection from the withering cold.[9]

STUDY 5-8

1 The characteristics of tundra vegetation—

in particular the absence of large plants—are mainly a result of the following factors. Comment on the influence of these controls on tundra vegetation:

   a summer temperature
   b length of growing season
   c drainage
   d permafrost
   e soils
   f length of daylight

2 Within a small area of the tundra minute variations in slope and drainage cause sharp contrasts in plant types. Explain why slope and drainage assume more importance as controls of vegetation in the tundra than in other regions.

3 The sequence of plant communities from the base of a mountain to its summit corresponds to the sequence from the equator to the pole. In an atlas find Mount Ruwenzori (5000 m) in Uganda and decide what types of vegetation would occur on its slopes.

4 To what extent will generalizations about vegetation in the arctic tundra also be true of alpine tundra? Referring to page 107, point out some differences in climate and other conditions that might be expected to cause some differences in vegetation. Write a brief, general description of alpine vegetation.

[1]Nicholas Polunin, *Introduction to Plant Geography and Some Related Sciences* (New York: McGraw-Hill Book Co., Inc., 1960), p. 3. Copyright 1960 Nicholas Polunin. Used with permission of McGraw-Hill Book Co., Inc.

[2]Vernor C. Finch, *et al.*, *Elements of Geography* (4th ed.;

New York: McGraw-Hill Book Co., Inc., 1951), p. 409.
Copyright 1957 McGraw-Hill Book Co., Inc. Used with
permission of McGraw-Hill Book Co., Inc.

[3]Source unknown.

[4]Finch, *Elements of Geography*, p. 409.

[5]Preston E. James and Clarence F. Jones, eds.,
American Geography: Inventory and Prospect (Syracuse, N.Y.: Syracuse University Press, 1954).

[6]A.M.A. Aubreville, 'Tropical Africa', *A World Geography of Forest Resources* (New York: The American Geographical Society, Ronald Press, 1956), pp. 361-5.

[7]Polunin, *Introduction to Plant Geography and Some Related Sciences*, p. 337. Copyright 1960 Nicholas Polunin. Used with permission of McGraw-Hill Book Co., Inc.

[8]S.R. Eyre, *Vegetation and Soils* (London: Edward Arnold, 1963), pp. 47, 49, and 50.

[9]W. Ley and the editors of *Life* magazine, 'The Poles', *Life Nature Library* (New York: Time Incorporated, 1962), p. 108.

6/ SOILS

Many city-dwellers see little soil from one year's end to another, yet soil is essential to life as we know it. This layer of material, formed over much of the earth's land surface, varies in depth from a few centimetres to several metres. Soil supports most plant life, including those plants necessary to human existence. Unlike variations in natural vegetation, which are easily recognizable, variations in soil cause almost no change in the observable character of some parts of the earth's surface. The different characteristics of soil types do however exert great influence on agriculture, which, in turn, makes possible almost all other forms of human activity.

To understand the differences between soils, including their various influences on vegetation and agriculture, it is necessary to have some knowledge of the general characteristics of soil, the factors responsible for the formation of these characteristics, and the nature and distribution of different types of soil. These three topics form the basis of this chapter.

THE GENERAL CHARACTERISTICS OF SOIL

The word soil refers to the relatively thin top layer of the earth's surface materials. These materials are usually found in distinctive layers or *horizons*. Soil consists of both mineral particles (weathered rock) and of organic particles (animal and vegetable material) which support the growth of plants. The subsoil beneath is almost entirely mineral. Soil can be generally defined as the medium for plant growth because it supplies the nutrients (including water and oxygen) for plant development as well as providing an anchor for the root systems of plants.

Composition

The larger part of the soil's bulk consists of mineral matter derived from the weathering of rocks. Weathering (described more fully on pp. 192 to 195) involves the breaking up of the solid rock of the earth's crust into loose material known as *regolith*. Regolith is not soil, but it does include the mineral particles that can be transformed into soil. The various minerals that make up the regolith are found in the soil in approximately the same proportions (see Fig. 7-4) as they are found in the regolith. Almost 90 per cent of the material in the earth's crust consists of minerals formed from combinations of the elements oxygen, silicon, aluminum, and iron. Other elements—such as carbon, nitrogen, phosphorous, calcium, and potassium—occur in smaller quantities but are of great importance in the development of plants.

The organic material in the soil comes from a number of sources. It includes the decaying leaves, stems, and roots of plants, decaying animal matter, and also living and dead micro-organisms. Following the death of any living organism, plant or animal, its organic compounds are eventually transformed by means of the decomposer chain (p. 17) into a few simple chemical products. These include oxygen, carbon dioxide, nitrogen, and various mineral salts. Such chemicals are among the nutrients necessary for the development of plant growth. Before decay is complete, the blackish organic matter is called *humus*. In addition to being an important source of nitrogen and phosphorus, this dark-coloured material also enhances the soil's capacity for absorbing and retaining both moisture and soluble plant foods.

In a good quality soil about 50 per cent of its volume consists of water and air, both of which are necessary to the many chemical reactions that occur in the soil. Without water, plants cannot absorb nourishment and the soil cannot support life. (All of the elements used by plants are dissolved by water in the soil and taken in through the roots.) However, few plants can exist where water excludes air from all the spaces between soil particles, even though air in the soil does not provide plant food.

Various living organisms inhabit the soil and are involved in its formation. They range in size from minute bacteria involved in the processes of decay to larger organisms such as earthworms and insects. These aid in the aeration of soil and the movement of water within it.

Texture and Structure

The texture of a soil is determined by the proportions within it of individual particles of different size. These particles fall into one of three classes: sand (over 0.05 mm in diameter); intermediate particles known as silt; and very fine grains of clay (less than 0.002 mm in diameter). There are many variations in texture resulting from combinations of these different-sized particles (Fig. 6-1). For example, *loam* is the term used for soil containing moderate amounts of clay, silt, and sand.

The size of the particles affect the movement and retention of water and, to a lesser extent, air. The larger particles, such as sand, permit water to pass through the soil very freely. Sandy soil, therefore, will dry out quickly after rain. Soil made up largely of clay, the smallest particles, is usually impervious to water and air. Thus in humid regions the spaces between the clay particles near the surface of the soil may be filled with water for substantial periods. The resulting deficiency

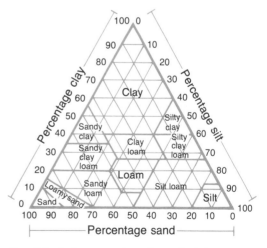

Fig. 6-1. Soil texture. In order to find the textural class of a soil, the percentages of clay, sand, and silt are entered on the appropriate scales, and the lines for each are followed to their intersection.

of air in this type of soil will inhibit plant development. In addition, soil containing a large proportion of clay is nearly always difficult to cultivate.

The structure of soil itself—the grouping of individual particles into larger, fairly regular pieces—is another important factor in determining the amount of space between particles in a soil. Occasionally soil has no structure in that the particles cling together in large irregular masses, or each grain is separate. This often occurs in the case of loose sand.

Generally speaking, a good soil has enough sand to promote good drainage, enough clay to promote aggregation and reasonable water holding capacity, and sufficient silt to give body without promoting either excessive drainage or imperviousness.[1]

The best soil structure for plants is a granular one in which pore space—the space between particles that makes a soil porous—amounts to 35-50 per cent of the soil's volume. The maintenance of such a structure is one of the main objects of good soil management. Unlike its texture, the structure of a soil can change either naturally, as the amount of water in the soil varies, or as a result of cultivation. Cultivation may crush the granules, particularly when the soil is very wet or very dry. This will in turn destroy the structure of the soil by causing the particles to run together. This condition makes the cultivation of plants almost impossible. A structureless soil, even though it is fertile, may need several years of careful treatment before it can be made productive again.

Colour

The colour of soils varies considerably. It can often be used as a general indicator of a soil's physical and chemical conditions. Soils of red, brown, or yellow colour, common in tropical and sub-tropical regions, usually indicate a concentration of the oxides of iron and aluminum or, more significantly, the absence of other minerals necessary for plant development. White or light-coloured soils, usually found in dry areas, may indicate an absence of iron and aluminum oxides or the presence of a concentration of soluble salts. The dark-brown or black-coloured soils found in grasslands usually denote large amounts of organic material. Colour is determined not only by mineral content but also by the amount of water in the soil, and great variations in colour within a small area may be the result of variations in drainage. The assumption that dark soils are fertile and light-coloured soils less fertile is often correct—but there are many exceptions.

Horizons

The character and composition of a soil can be observed when an undisturbed area of the top layer of the earth's surface is cut into a depth of several decimetres. This exposes a *profile* of the soil from the surface down to the underlying parent material. Vertical differences in composition, texture, structure, and colour appear at varying depths, dividing the soil into zones known as *horizons* (Fig. 6-4). A skilled observer can determine quite accurately the characteristics of a soil by examining the horizons exposed in a profile.

A study of a number of soil profiles will show that there are great variations in soil characteristics from place to place. Before attempting to study the nature of these variations, it is necessary to understand the processes that cause them and form the different types of soil. These are discussed in the following section.

STUDY 6-1

1 What are the different substances that compose a soil? Explain the importance of each of these to the development of plants.

2 Referring to Fig. 6-1, identify the textural

class of soil to which each of the following examples belongs:

	sand	clay	silt
a	65% sand	15% clay	20% silt
b	34% sand	33% clay	34% silt
c	40% sand	18% clay	42% silt
d	10% sand	45% clay	45% silt
e	17% sand	13% clay	70% silt

3 Explain the importance of texture and structure in determining the fertility of a soil.

THE FACTORS IN SOIL FORMATION

While soil may appear to be a lifeless, unchanging layer, it is actually a dynamic, ever-changing zone in which complex biological, physical, and chemical activities are constantly occurring.

Parent Material

The parent material of soil is derived from the breakup of the outer rock of the earth's crust. The fragments of rock that result from various weathering processes (see p. 192) are subject, at the top of the regolith, to further processes that result in the formation of soil. Before they become part of the soil such rock fragments are termed parent material. The material, of course, varies according to its origins. It may consist of rock materials formed from the underlying bedrock by weathering, or it may have been transported from its original location by running water, wind, or moving ice. The latter materials will bear no relationship to the underlying bedrock over which they have been deposited.

The degree to which parent material has influenced the characteristics of a soil usually depends on the age of the soil in question. As soil ages the characteristics it inherits from parent material may be so completely modified by other processes that it will bear little or no resemblance to the parent material. Where some resemblance persists, it is usually in the relationship between the mineral composition of the parent material and the soil. For example, when such minerals as quartz, iron and aluminum hydroxide, and clay occur in large quantities in the parent material, they will usually also be found in the soil because of their resistance to the soil-forming processes. In other instances, where certain elements necessary for plant growth are missing from the parent material, they are likely to be missing from the soil as well. This deficiency may limit the soil's fertility.

Climate

Both precipitation and temperature are important factors influencing the transformation of parent material into soil.

Water that percolates downwards through decaying organic matter and weathered particles carries soluble minerals towards the ground water zone (see p. 26). This process is known as *leaching*. It varies according to the amount and effectiveness of precipitation and the porosity of the parent material. Moderate leaching is normal and necessary for the development of soil. If leaching is excessive—as is often the case in very humid regions—the soil loses many of the nutrients for plant growth. The deposition of the leached material may occur within the body of the soil itself, or it may occur much farther down, below the body of the soil and beyond the reach of most plant roots. In addition to transporting minerals in solution, the percolation of water also causes the downward movement of fine, solid particles into the lower part of the soil. Coarser particles are left near the surface.

On the other hand, the horizontal movement of water across the top of the soil will cause erosion and the loss of some top soil. Under normal circumstances a small loss is not critical since it can be replaced by the parent material and humus. Too much erosion, however, can be damaging because it results in the removal of more top soil that can be replaced through natural processes.

(This problem is often created when the protective cover of vegetation is destroyed.) Little or no horizontal movement of water on the top of the soil may result in either excessive leaching or, in flat low-lying areas, poor natural drainage and a lack of air in the soil.

In sub-humid and arid areas, potential evaporation exceeds precipitation. As a result the surface of the ground is often much drier than the soil and the parent material beneath it. Given these conditions, the water is held in the pore spaces, as loose film around the grains of soil, and even in the parent material. Eventually it will move upwards—a short distance at least—towards the drier surface. This process is known as *capillary action*. This movement occurs in the same way that fuel is drawn upwards through a wick in an oil lamp. A particularly fertile soil may often result from this process because soluble minerals (particularly calcium carbonate) are brought closer to the surface.

It is only in very arid regions, particularly where there is irrigation, that capillary action may result in an excessive accumulation of soluble salts at or near the surface. This produces an alkaline soil that will harm or even prevent the growth of many plants.

Temperature is another important factor influencing the formation of soil. It affects both the speed and the nature of the chemical reactions taking place in the soil's inorganic material. It also influences the rate of bacterial activity in the soil's organic matter. In tropical areas particularly, high temperatures and excessive precipitation can transform parent material into soil that bears little resemblance to the bedrock from which it originated. The same tropical conditions also promote the very rapid decay of organic matter into its constituent elements and their subsequent leaching far beneath the surface. Conversely, in cold areas low temperatures retard many of these processes, and chemical changes and bacterial activity occur very slowly.

Plants and Animals

Living plant and animal micro-organisms form a small (1-2 per cent) but important part of soil's organic content. The plant group includes bacteria, actinomycetes, and fungi. Protozoa (one-celled organisms) and nematodes (microscopic worms) are the principal members of the animal population. These organisms are important for plant development because they affect the soil by absorbing nitrogen from the air and adding it to the chemical content of the soil. These minute organisms, which occur in vast numbers, also enrich the soil's organic content when they die. Larger living creatures, such as earthworms and burrowing animals, assist on a smaller scale in the process of soil formation. They mix materials and bring subsoil minerals to the surface, and, when they die, add their own remains to the soil.

While micro-organisms form part of soil's organic content, most of the organic matter in the soil comes from the decomposition of plants. A plant takes water and nutrients from the soil and air. When it dies these are broken down by billions of micro-organisms and incorporated in the soil. Deep-rooted plants bring mineral solutions up from below the soil and build them into their tissues; when such plants die these minerals are added to the soil. It is during the process of decomposition, as we have seen, that valuable humus is formed.

The amount of humus in the soil depends on the type of vegetation, the rate of decay, and the amount of leaching. For example, soils formed under grasslands are enriched every year by a large supply of dead leaves and roots. In middle-latitude grasslands this material decays at a moderate rate and, because of the fairly low annual precipitation, is not severely leached. In contrast to grassland soils, soils formed under forests are not as enriched by humus. In needleleaf evergreen forests the rate of decay is slow, and a spongy layer of partially decomposed organic material covers the surface. This layer is highly acidic. It causes both organic and inorganic soluble material to be leached from the soil. In tropical forests extremely rapid decomposition is accompanied by such excessive leaching that little organic material can accumulate in the soil.

Surface Configuration

As already noted, the rate of erosion affects the depth of the soil while the slope of the land influences the amount of moisture it contains. Soils on well-drained, gently sloping or rolling land are usually the best developed because the losses that result from natural erosion are balanced by the downward progress of the soil-forming processes. These processes are retarded in steeply sloping areas by increased erosion and insufficient leaching. In low-lying level areas soil development may also be curtailed if leaching is restricted and if air, which is essential to the chemical processes, is absent from the soil.

Human Influence

Soil is created by the interaction, over a long period of time, of parent material, climate, plants, animals, and surface configuration. In areas little influenced by human development, such as forest, desert, and tundra regions, the soil is entirely a product of these natural processes. However in areas where the soil has been used to produce food and other materials, its natural characteristics have been altered, in some cases quite substantially. Good soil management is necessary if soil is to be used to produce crops year after year without any loss in fertility. Some soils in parts of southern and eastern Asia have been continuously cultivated for thousands of years. Indeed, people are capable not only of maintaining soil fertility but also making infertile soils productive. It also fol-

lows that poor management, which occurs in many places, can ruin soils. The reclamation of damaged soils may require either many fallow years to permit natural regeneration or considerable expenditure on fertilizers and other special measures.

There are several different ways in which human activity can have an impact on soil. First, the role of natural vegetation in the soil-forming process is altered when such vegetation is cut down or grazed by animals. One effect of this type of change is to reduce the organic content of the soil. A second form of human impact occurs when cultivation takes place. Constant ploughing and cropping alter a soil's structure, remove some of these elements that crops require for growth, and usually increase the rate of surface erosion. Finally, continuous soil use requires good soil management practices. Such practices include: removing excess moisture by digging drainage ditches; increasing moisture through irrigation; maintaining or improving the fertility of the soil by the addition of mineral and organic fertilizers; and using various farming practices such as crop rotation that are intended to maintain or enhance soil fertility.

Plant Nutrients

Plant tissue is composed of approximately 45 per cent carbon, 43 per cent oxygen, and 6 per cent hydrogen. These elements are derived from the atmosphere (carbon dioxide and oxygen) and water. At least sixteen other elements make up the remaining 6 per cent. Most of these elements are necessary for plant growth and are found chiefly in the soil. Such elements are classified as macronutrients and micronutrients, depending on the quantity of each in plants.

The principal macronutrients are *nitrogen*, *phosphorus*, and *potash* (potassium). They are essential for plant growth. The bulk of most commercial fertilizers is made up of

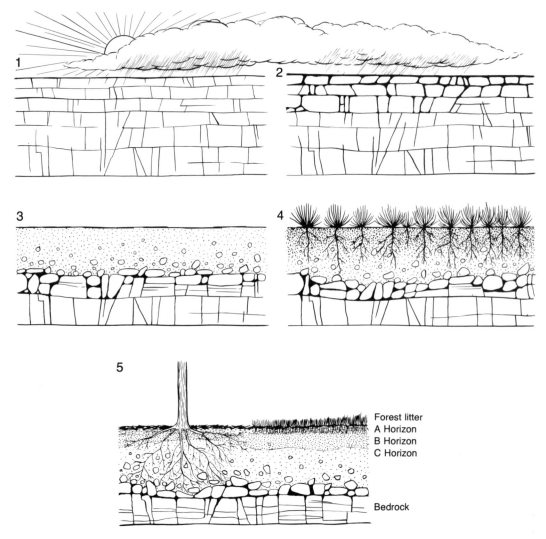

Forest litter
A Horizon
B Horizon
C Horizon

Bedrock

Fig. 6-2. These five stages represent the development of a soil. In the first three, the work of weathering slowly causes the bedrock to be broken up and a layer of regolith is established. This becomes the parent material. By the fourth stage pioneer plants are established. Their remains, together with that of the years of plant growth following them, add organic material and new minerals to the soil. Finally, by the fifth stage, a mature profile with well-marked horizons has developed.

these nutrients. Their relative quantity is indicated on a bag of fertilizer by three numbers —for example, 6-24-24 means a minimum of 6 per cent nitrogen, 24 per cent phosphorus, and 24 per cent potash. Nitrogen is a major component of plant proteins and a deficiency of this nutrient in the soil probably limits plant growth (particularly leaves and stems) more than any other factor. Phosphorus promotes vigorous growth (particularly root and seed development). Potassium increases resistance to disease. Other macronutrients include calcium, magnesium, and sulphur.

Micronutrients are just as essential to plant development as macronutrients but they are needed in much smaller quantities. A deficiency of these elements in the soil is generally caused by the same things that cause a deficiency of macronutrients—leaching, erosion, and removal by crops. Some of the more important micronutrients include boron, copper, iron, manganese, and zinc.

STUDY 6-2

1 Figure 6-2 illustrates a possible sequence in the development of a type of soil known as podzol. It shows its development from an exposed surface to a mature profile.

a Explain the processes involved from Stage 1 to Stage 3. In what ways will slope and variations in climate influence this development?

b The first or pioneer plants use the nutrients provided by weathered rock, taking from it nitrogen, soluble minerals, and water. The carbon and oxygen required by the plants come from the air. When these plants die, how do they contribute to the formation of the soil? As this cycle of growth and decay occurs again and again, the fertility of the soil increases. Explain.

c What determines the stratification of the various materials in the horizons? Why is

the final stage described as a mature soil profile?

2 Soil fertility, or the availability of plant nutrients, has been mentioned in this chapter a number of times. In general, fertility means the ability of the soil to support plants for human use or consumption.

a The single most important factor affecting fertility is the soil's ability to provide mineral nutrients. What causes variations in the mineral content of different soils?

b What other natural factors will affect the fertility of the soil?

c The fertility of soils has been reduced by careless farming practices such as poor crop selection, overcropping, incorrect ploughing, and improper irrigation. Excess soil erosion is often a direct result of the destruction of vegetation on sloping lands or in dry areas. Explain how all of these practices lead to a decline in the productivity of soil.

d It is inevitable that some rather major changes will result from our use of the soil. These changes, however, need not cause a decline in the soil's productivity. Measures such as those listed below will, when applied properly, maintain and even increase productivity. Explain how these measures will help to maintain the soil and note the conditions under which each would be used:

i the use of mineral fertilizers

ii fallowing and rotation

iii manuring

iv dry farming

v water control (irrigation or drainage, or erosion control).

e Soil fertility is both a biological and an economic phenomenon. Explain this statement.

f Discuss the validity of the following two statements.

i In terms of human existence, soil is as important as air and water.

ii Something difficult to replace must be

disappearing or being misused before conservation is necessary.

A CLASSIFICATION OF SOILS

In order to study the geography of soils, some system of classification must be used. Like all classifications of physical phenomena, however, this requires a number of generalizations.

In the system of classification outlined below, *soil orders* are the first level. There are three distinct soil orders: zonal, intrazonal,

and azonal. Zonal soils are mature soils developed on well-drained land and reflecting definite climatic influences. Intrazonal soils differ from the zonal because their development has usually been influenced by some local soil-forming factors other than climate or vegetation. These may include poor drainage or an unusual parent material. Azonal soils are those composed principally of sand or alluvium. Because of their recent formation, these soils exhibit little or no profile development.

The soil orders are divided into *sub-orders* according to major climatic and vegetational

System of Soil Classification

Order	Sub-order	Great Soil Group
	Soils of the cold zone.	Tundra
	Light coloured soils of arid regions.	Desert Red desert Sierozem
Zonal	Dark coloured soils of the semi-arid, sub-humid, and humid grasslands.	Brown Chestnut (Dark brown) Chernozem
	Soils of the forest-grassland transition.	Degraded chernozem Prairie
Zonal	Light coloured podzolized soils of forested regions.	Podzol Brown podzolic Grey-brown podzolic Red and yellow podzolic
	Soils of warm sub-tropical and tropical regions.	Latosols
	Saline soils of poorly drained arid regions.	Saline
Intrazonal	Poorly drained soils of humid regions.	Meadow Bog Hardpan
	High lime soils.	Rendzina
Azonal		Lithosols (rocky soils) Alluvium Dry sands

differences. These sub-orders in turn are further divided into *great soil groups*. The important differences between the great soil groups are again a result of variations in temperature, precipitation, and vegetation. These groups are used to describe the world pattern of soils (Fig. 6-3). Soil scientists further divide each great soil group into families, the families into series, and the series into types. The result is a classification consisting of thousands of soil types. For the purposes of this book, it is enough to examine some of the more important great soil groups.

The Great Soil Groups

The following material, together with Fig. 6-3 and Studies 6-3 through 6-7, will permit a general examination of some of the more important great soil groups. The object of this examination is to determine the characteristics and world distribution of these groups. From examining examples of specific soils in this section it is possible to come to some understanding of the characteristics of other soil types in the same sub-order.

PODZOL SOILS

The podzol soil group is found over large areas in the higher middle latitudes of the northern hemisphere. It is developed under the cool summer microthermal and sub-arctic climates and under needleleaf vegetation. The profile (Fig. 6-4) shows a number of fairly well-marked horizons. At the top is the A horizon. A_1 horizon is normally composed of a spongy layer of leaf mould or badly decomposed organic remains. This strongly acidic layer is sharply separated from the A_2 horizon —the eluviation zone. Here, the acid solution has leached away the aluminum and iron, together with most of the clay and organic particles, leaving a bleached and almost structureless ash-grey layer (the word podzol is derived from Russian words mean-

ing 'ash soil'). The B horizon is usually much darker in colour, having been enriched by some of the leached material from above. In many areas the C horizon is composed of glacial drift.

STUDY 6-3

1 Make a sketch of the podzol profile shown in Fig. 6-4, indicating and identifying the different horizons.

2 Generally podzol soils are not very fertile. They are notably deficient in calcium, magnesium, potassium, and phosphorus. (Conifers need only small amounts of these nutrients.) What are some of the reasons for the infertility of podzol soils?

GREY-BROWN AND BROWN PODZOLIC SOILS

The grey-brown and brown soils of middle-latitude forest regions are known as podzolic soils because they have been formed by processes similar to those responsible for the podzol group. Unlike podzol soils, however, they are among the more productive agricultural soils. There are several reasons for this. Leaching, the major soil-forming process, is not as severe in them as it is in podzols farther north. The uppermost horizon of the soil is composed of partly decayed organic material two to seven centimetres deep. Owing to a different vegetation and warmer temperatures, this layer is well decomposed and not an acidic, matted layer like the A horizon in the podzol. In addition, the organic material derived from broadleaf forests contains more lime and potash than does that from needleleaf forests, and these are easily mixed with the soil's minerals by earthworms and other soil organisms. In most instances the structure of the soil is suitable for cultivation and responds readily to the application of fertilizers.

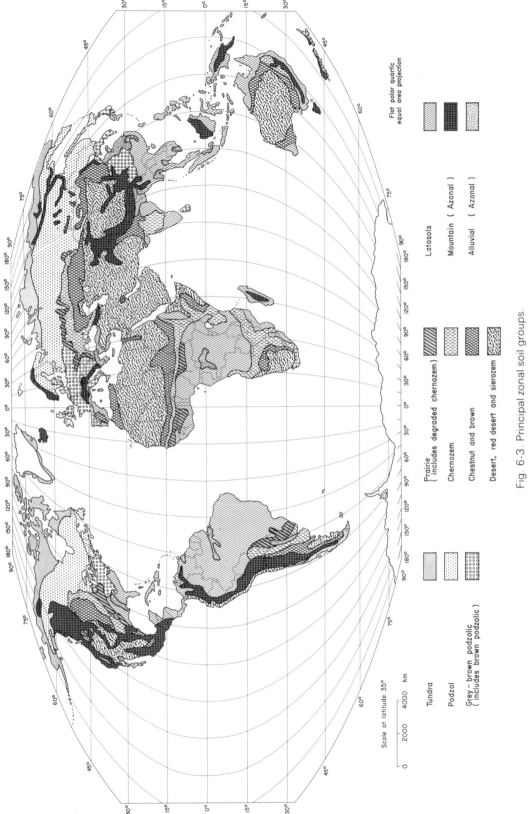

Fig. 6-3. Principal zonal soil groups.

Scale at latitude 35°

0   2000   4000   km

Tundra

Podzol

Grey – brown podzolic
(includes brown podzolic)

Prairie
(includes degraded chernozem)

Chernozem

Chestnut and brown

Desert, red desert and sierozem

Latosols

Mountain (Azonal)

Alluvial (Azonal)

Flat polar quartic
equal area projection

Fig. 6-4. Profile of a podzal formed from glacial till in Kamouraska County, Quebec. The scale is marked in inches. (The top of the profile can be seen to have an A₁ and an A₂ horizon.)

STUDY 6-4

1 Describe the distribution of both the podzol and the grey-brown and brown podzolic groups, referring to latitudinal extent as well as general location.

2 Make a sketch of the profile shown in Fig. 6-5, indicating and identifying the different horizons mentioned in the caption.

3 Describe the influence of climate and vegetation on the formation of this soil type and compare it with a podzol soil.

Fig. 6-5. Profile of a grey-brown podzolic formed from glacial till in southwestern Ontario. The A_1 horizon is grey-brown with a high content of well mixed organic matter. The A_2 horizon, like the podzol, is leached, although not to the same extent. This horizon is normally light brown or brownish yellow in colour. The B horizon is much darker in colour with more clay than the A horizons and has an irregular, block-like structure. The C horizon, or parent material, is usually grey, often stony, and contains the calcareous material leached from the upper horizons.

LATOSOLS

The term latosol does not refer to a single type of soil. It refers to a group of soils found on well-drained sites in the humid tropics. While subdivisions of this group can be made, the following material describes some of the more important features that tropical soils have in common.

Tropical soils are poorer and more fragile than those of temperate regions. Great care is needed in using them if their further impoverishment and destruction are to be avoided. These conditions give tropical agriculture a precarious character which is absent from the temperate belt, except in sub-arid regions. . . .Why are tropical soils, with rare exceptions, so infertile? They are often deep, for the bedrock may have decomposed to a depth of several scores of feet. But the soluble matter, bases and nitrates, is soon carried away by percolating water aided by the high temperature the presence of carbonic acid and nitric acid, and by countless bacteria. . . .Tropical soils are poor in humus. Even in the forest humus blackens the soil to a depth of only a few inches. In dense forests the quantity of organic matter deposited is considerable and has been estimated in Yangambi in the Congo at between twenty and twenty-five tons of leaves, twigs, lianas, and branches per annum per acre. But in these forests the deposits do not enrich the soil, for they are offset by equivalent losses, with the result that the soil contains at most 1.8 per cent of humus, while fertile soil in temperate regions often contains more than 10 per cent.[2]

Tropical soils have several other important characteristics. They are granular, porous, and very permeable (water can easily flow through them). Owing to these characteristics, they may be tilled immediately after heavy rains, but these same qualities make them subject to drought. Because they are highly leached, many tropical soils—especially latosols—are low in both mineral and organic plant foods. Consequently they require fertilization in order to produce the same quantity of crops year after year.

Leaching and bacterial action remove most of the organic materials and soluble minerals from this soil. These processes leave oxides of aluminum and iron (not very soluble in mildly acidic soil water) that give the soil its typical reddish colour. In some localities these iron and aluminum minerals form material (not a soil) called *laterite*. When dried, it is often used as a building material. Valuable mineral deposits also occur as laterites: bauxite (hydrous aluminum oxide), limonite (hydrous iron oxide) and manganite (manganese oxide).

Soils formed under tropical rainforests are difficult to farm continuously. One traditional farming system known as shifting cultivation does this successfully by using the natural ecosystem to replenish the supply of nutrients and organic material in the soil. There are many variations of shifting cultivation but most involve clearing a plot of the land, cultivating it for several seasons, and then allowing it to return to forest. The cultivator then clears a new section of forest land. The soil in the original plot of land, after many years of natural regeneration, will be replenished. Eventually it can be farmed again.

Unfortunately, as population densities have increased in many tropical countries, it has been necessary to shorten the length of the period of fallow. As a consequence crop yields have become lower and the soil structure has deteriorated. The maintenance of tropical soils in permanent production is a difficult task. One approach to solving this problem is establishing a system of land use that is similar to the natural forest ecosystem. A land use system based largely on tree crops is an example of such a system. Permanent pasture grasses growing among the trees would reduce soil erosion and at the same time could be used to feed livestock, thus diversifying agriculture in tropical regions.

1 Describe the influence of climate and vegetation in the formation of this soil type. Within the latosol group there are a great many variations. What are the reasons for these variations, and why is this group not divided further?

2 As illustrated in Fig. 6-6, latosol profiles are normally very deep and seldom show well-marked horizons. What are the reasons for these characteristics?

3 What processes are responsible for the inadequate quantities of plant nutrients in this soil group?

4 If tropical soils are so infertile, how can they support such abundant natural vegetation?

5 What environmental assets for plant growth do tropical regions possess? The possibilities of increasing the amount of agricultural land in these areas are very great. What factors, other than natural, have inhibited agricultural development in the tropics? How might the existing situation be improved, and what would be the effects of such improvements on the life of the people?

CHERNOZEM SOILS

Chernozem is a Russian word meaning 'black earth'. These fertile soils are particularly important in the sub-humid prairies of the United States, Canada, the Ukraine, and Argentina. They are formed under dense grasses and in conditions of low annual precipitation.

The number of crops that can be grown without irrigation is limited in such regions because the summers are hot and dry and there is a deficiency of soil moisture. Chernozem soils are often cultivated for wheat or barely—crops that can survive under these conditions and produce grain of high quality. Much of the world's surplus or exportable food comes from these 'breadbaskets'. The

Fig. 6-6. Profile of a latosol formed of material weathered from gneiss, north of Rio de Janeiro, Brazil. This has the characteristic, darkened A horizon to a depth of approximately 30 cm, the absence of evident horizons below, and the normal penetration of plant roots to great depths. Numbers on the scale indicate feet.

existence of this surplus is particularly important in years when some parts of the world experience food deficiencies.

1 The average chernozem profile is made up of three easily distinguished horizons. The A horizon, a black layer rich in humus, has a granular structure. The B horizon is lighter brown in colour. It is usually quite easy to distinguish from the much-lighter-coloured C horizon.

    a Make a sketch of the chernozem profile shown in Fig. 6-7, showing the three horizons. Using Fig. 6-7, and your sketch of it, summarize the main soil-forming processes responsible for the development of chernozem soils.

    b The white spots in the lower part of the profile are accumulations of calcium carbonate. What has caused their formation?

2 Compare the location and characteristics of the chernozem and brown soil groups. Explain the differences as well as the similarities between them.

CHESTNUT AND BROWN SOILS

This group is found on the dry margins of the chernozem soil and also under grassland. It is named for the colour of its A horizon, which is a result of the accumulations of dead organic matter. Grasslands supply less organic material than most forests, and the rate of decay of these materials is slower. Since precipitation is generally low, the amount of leaching is small, and soluble minerals (principally calcium carbonate) are seldom carried more than 35 to 50 cm below the surface. The zone of illuviation so formed is often called the horizon of lime accumulation. The soluble minerals in this horizon have another source: the evaporated minerals that are brought upwards into the soil profile by capillary action.

Fig. 6-7. Profile of a chernozem soil formed from glacial till in east central South Dakota. Numbers on the scale indicate feet.

On the average probably less than 40 per cent of these soils are cultivated. Their cultivation occurs mainly in areas where the type of soil material (e.g. clay) or the topography is favourable to water retention. Much of the uncultivated chestnut and brown soils are covered with grasslands that are often used for pasture. Wind erosion is a serious hazard if the organic content of these soils is depleted by farming. It is intensified in years of below-average precipitation. The dust-storms associated with the prairies during the 1930s are an example of this phenomenon. Since then, various conservation practices have been used to preserve moisture and also reduce soil losses from wind erosion. These include summer fallowing, restricted grazing, and special ploughing techniques. Brown and chestnut soils have a reasonably high degree of natural fertility. They are very productive during years of above-average pre-cipitation, or when they are irrigated.

STUDY 6-6

1 Where are the principal areas in which the brown soil group is found? What soil groups appear on its arid and humid margins?
2 Describe the influence of climate and vege-tation in the formation of this type of soil.
3 Why do the brown soils have a shallow profile? What causes the horizon of lime accu-mulation? At what depth is it found? Why is it nearer the surface of the brown soil group than of other grassland soils?
4 If the brown soils have a reasonably high degree of natural fertility, why is only 25 per cent of the area of brown soils on the Cana-dian prairie classified as arable land?
5 Why did severe dust storms occur on the prairies during the 1930s? What is being done to prevent this from happening again? Under what conditions might the area of ara-ble land be increased?
6 Desert soils are characterized by very shal-low profiles, a lack of organic material, accu-mulations of calcium carbonate at or near the surface, and a pale colour ranging from red (tropical deserts) to grey (mid-latitude deserts). Explain these characteristics and compare them to similar characteristics of the chestnut and brown soils.

ALLUVIUM

Alluvial soil probably provides food for a larger percentage of the world's population than any other type of soil. The composition of alluvial soils, which are found in river valleys and coastal regions, varies according to the nature of the material deposited by rivers. Most alluvial soil contains a good supply of plant nutrients and is generally free of stones and easy to cultivate—but in many areas it is poorly drained. The periodic flooding that occurs on some floodplains adds minerals and humus to the soil. It is only when such deposits no longer occur that the soil-forming processes begin to turn alluvial material into a zonal soil.

STUDY 6-7

1 Describe the importance of this soil group by examining the distribution of alluvial soils (Fig. 6-3) and comparing their distribution with the distribution of population. It should be noted that there are many areas of alluvial soil that are locally important but are too small to be shown on a world map.
2 Although alluvial soils are normally fertile, there are often many problems associated with their use. What are some of these?

The 'Seventh Approximation' or Soil Taxonomy

The classification of soils discussed in this chapter has proved very durable. It has been in use in North America for many years and remains a suitable framework for describing and interpreting the general pattern of the world's soils. It has proved less useful for the objective and quantitative classification of the soils of smaller areas such as a county or even a farm. To overcome deficiencies in the system, soil scientists in the United States developed a new scheme during the 1950s and early 1960s. This new classification,

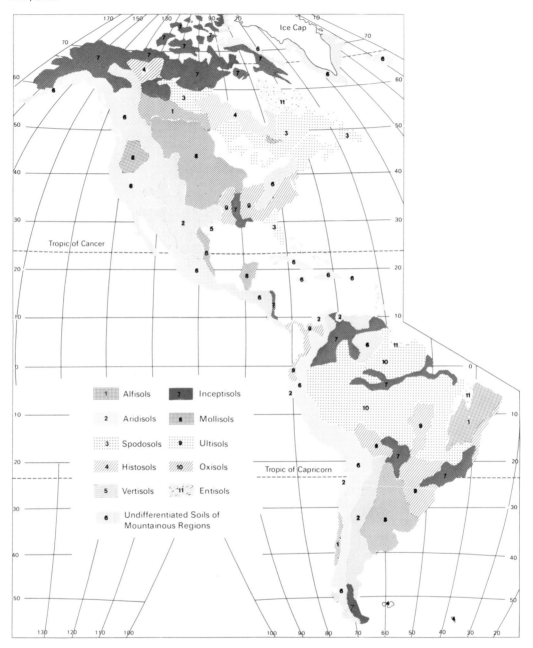

Fig. 6-8. This is a highly generalized map of the soil orders of the soil taxonomy. It is based on a map prepared by the Soil Conservation Service of the U.S. Department of Agriculture.

Name	Definition	Horizons	Important Characteristics Nutrients and Fertility
HISTOSOLS	Greek *histos*, tissue	High concentrations of organic matter in upper 80 cm.	In cool climates, acid reaction, low in plant nutrients. In low and middle latitudes highly productive when treated
SPODOSOLS	Greek *spodos*, wood ash	Reddish 'B' horizon mainly from compounds of aluminium and iron.	Strongly acid. Low in nutrients.
OXISOLS	'oxi' short for oxide	Lack distinct horizons in low latitudes. Usually red, yellow, and yellow-brown.	Extreme weathering of most minerals. Supplies of nutrients commonly low.
VERTISOLS	Latin *verto*, to turn	Horizons not usually definite.	Intermediate amounts of organic matter. Usually neutral in reaction.
ARIDISOLS	Latin *aridus*, dry	One or more horizons. Surface horizons not normally darkened by humus.	Carbonate accumulations at depth. Often very productive under irrigation.
ULTISOLS	Latin *ultimus*, ultimate	Accumulation of clay in B horizon. B horizon often red or yellow.	Availability of calcium, magnesium and potassium highest in top 10 cm but diminishes rapidly with depth.
MOLLISOLS	Latin *mollis*, soft	Dark brown to black surface horizon.	One of most naturally fertile soils in the world
ALFISOLS	*al* and *fe* chemical symbols for aluminium and iron	Gray, brownish or reddish upper horizon, not darkened by humus.	Medium to high availability of calcium magnesium and potassium
INCEPTISOLS	Latin *inceptum*, beginning	One or more horizons present but little translocation of materials.	Considerable variation.
ENTISOLS	'ent' is associated with recent	Lack of readily recognisable horizons.	Variable fertility. Floodplains highly productive.

Fig. 6-9.

called Soil Taxonomy, became popularly known as the 'Seventh Approximation' after the seven successive attempts to devise an acceptable system.

The Soil Taxonomy has several characteristics that are desirable in any classification. Classes are defined in terms of actual or inherent characteristics of the soil that are measured quantitatively. Procedures for classification can thus be repeated with similar results by different soil scientists. No attempt is made to classify soils as they might have existed before human modification. Instead, they are classified in terms of their present properties. Six categories of soil are recognized by the Soil Taxonomy. The names of some categories are derived from the old classification. In the United States the different categories, and the number of classes in each, are as follows:

soil orders	—	10;
soil sub-orders	—	47;
great soil groups	—	180;
soil sub-groups	—	960;
soil families	—	4700;
soil series	—	11 000

THE SOIL ORDERS

At first sight the names given to the soil orders may seem complicated (see Fig. 6-9). They are new words, derived mainly from Latin and Greek, with a meaning to indicate a main characteristic of the soil. Alifsols (derived from *al*, aluminum, and *fe*, iron) have concentrations of these two elements; histosols (from the Greek *histos*, meaning tissue)

have high accumulations of organic matter. Once the names become familiar they are a useful reminder of each soil's basic characteristics. The main characteristics of the ten soil orders are presented in Fig. 6-9 and explored in Study 6-8.

STUDY 6-8

Figure 6-8, which shows the American distribution of the soil orders in the Soil Taxonomy, and Fig. 6-9, provide the necessary information for this study.

1 Name the approximate equivalents in the Soil Taxonomy for the following soils of the old classification: podzols, alluvial soils, and chernozems.

2 Histosols have a high content of oganic matter and are found in both high and low latitudes. Give reasons why large accumulations of organic mattter occur in each of these different climatic conditions.

3 Select four soil orders of the Soil Taxonomy, briefly relate their characteristics to the climate and vegetation of the areas where they are located, and comment on their agricultural productivity.

4 Why do the zonal, intrazonal, and azonal groups not exist in the Soil Taxonomy?

---

[1] E. Higbee, *American Agriculture* (New York: John Wiley & Sons, Inc., 1958), p. 34.

[2] P. Gourou and E.D. Laborde, *The Tropical World* (3rd ed.; London: John Wiley & Sons, Inc., 1961), pp. 13, 16, and 17.

Most people regard the earth's landforms as permanent features. No matter what other changes occur, we can feel secure with the mountains, hills, valleys, or plains that make up our local landscape. Sometimes we may detect a change in the course of a river or in the growth of small gullies on a hillside, or see in built-up areas impressive changes made by people, but most landforms remain unaltered during our lifetime. Pictures or maps made several decades ago reveal landforms essentially identical to those existing today.

Although we rarely see changes in the landforms of the earth, change is continually occurring. Since the formation of our planet (estimated at over 4.5 billion years ago) every part of its surface has undergone many alterations. For example, mountains exist today where once there were plains, and it is conceivable that the plains may have been formed from deposits of sediment that masked the roots of an even older mountain system worn down hundreds of millions of years ago. Before we can even begin to appreciate the rate of geological change, we need to discard normal concepts of time.

Man's short life is a poor yardstick with which to measure and understand the rates of earth-shaping events. It is like trying to express the distance from the earth to the moon in inches: the resulting numbers are so large they are meaningless. It is difficult to realize how much can be accomplished over millions of years of geologic time by the almost imperceptibly slow process that can be seen at work today.[1]

Little is known about the exact rate of geological change because, by human standards, changes occur incredibly slowly, and precise measurements go back only a few decades. Yet some indication of the present rate of geological change can be gained from the following observations. It has been determined, for example, that the coast of the Netherlands is sinking at the rate of 20 cm each century, and the city of Venice is sinking at the rate of approximately 30 cm a century. Two surveys of southern California made by the U.S. Coast and Geodetic Survey show that in one area the earth's crust rose 18 cm in three years and in another area 20 cm in 38 years. Further evidence can be found in southern California (where the earth's crust is notably unstable) where the horizontal shift along the San Andreas Fault is approximately 5 to 8 cm a year. The amount of shift has been sufficient to cause jogs in roads, rivers, and fence lines.

While little is known about the rate at which geological changes take place, a great deal is known about the events of the past

Fig. 7-1.
GEOLOGICAL TIME CHART[1]

GEOLOGIC DIVISIONS			MAJOR GEOLOGIC DEVELOPMENTS (MOST REFERENCES ARE TO NORTH AMERICA)	MAJOR BIOLOGIC DEVELOPMENTS
ERAS	PERIODS	EPOCHS		
CENOZOIC	Quaternary	Recent 10 000	Gradational activity developing present landscapes	Rise of civilization.
		Pleistocene 1 000 000	Four periods of glaciation. Tectonic activity form the coastal mountains of western North America.	Development of human beings. Extinction of larger mammals.
	Tertiary	Pliocene Miocene Oligocene Eocene Paleocene 63 000 000	Continents assume present form Marine invasion and deposition in present coastal regions Continued mountain building in the Rockies, Andes, and Himalayas	Emergence of early humans Development of horses, whales, monkeys, elephants, and large carnivores. Rise of grasses, cereals, and fruits.
MESOZOIC	Cretaceous	135 000 000	Mountain building towards the end of the period forms present major systems: Rockies, Alps, Andes, etc. (Laramide orogeny). Marine deposits.	First flowering plants. Extinction of dinosaurs.
	Jurassic	180 000 000	Mountain building by volcanic intrusions forms Selkirks, Cascades, etc. in western North America (Nevadan orogeny). Shallow seas cover much of the interior of North America and Europe.	First birds. Dinosaurs abundant.
	Triassic	230 000 000	Marine invasion and deposition. Widespread volcanic activity.	First mammals. First dinosaurs.
PALEOZOIC	Permian	280 000 000	Mountain building in eastern North America (Appalachian orogeny) and Europe (Hercynian orogeny). Ice Age in southern hemisphere.	Conifers abundant. Many insects, amphibians, and reptiles.
	Carboniferous { Pennsylvanian { Mississippian	345 000 000	Deposition of coal-bearing strata in eastern and central North America. Deposition of limey sediments in central North America.	First reptiles. Giant forest of (coal-forming) spore-bearing plants.
	Devonian	405 000 000	Mountain building in eastern North America (Acadian orogeny). Continued deposition.	First amphibians. Abundance of fish.

Silurian	452 000 000	Widespread marine invasion and deposition.	First appearance of plants and animals on land.
Ordovician	500 000 000	Some mountain building in New England (Taconic orogeny), mainly a period of continuing deposition.	First vertebrates.
Cambrian	600 000 000	Invasion of land by shallow seas, and deposition of marine sediments in geosynclines.	Abundance of marine invertebrates.
PRECAMBRIAN	Time extends back to the beginnings of the earth over 4.5 billion years ago.	The crust forms, and the continents and seas appear (although not in their present shape)	First known animal (jellyfish) 1.2 billion years. First known plant (algae) 3.2 billion years. Oldest known rock 3.3 billion years.

[1]Figures indicate the number of years before the present that an era, period, or epoch began.

and their effect on the crust of the earth. By piecing together various bits of information from a variety of sources, geologists have been able to reconstruct many of the changes that have occurred in the earth's rocks over several hundred million years. In order to put these changes into chronological perspective, geologists developed a time scale similar in principal to that used by historians. The main difference is that periods of human history cover tens or hundreds of years while periods of geologic or earth history are graduated in tens or hundreds of *millions* of years. The geological time chart shown below should be used for reference whenever geological or biological events of the past are mentioned.

We have noted that landforms are not permanent. At any given time they represent one stage in a continuing series of changes. The forces responsible for change are commonly divided into two major groups: *tectonic forces* and *gradational forces*. The two forces are closely linked. Tectonic forces derive their energy from sources within the earth and tend to alter its crust by producing irregulari-

ties. Gradational forces originate beyond the earth and draw energy mainly from the sun. They work continuously on the crust and tend to wear down irregularities that are produced tectonically. In the process they sculpture most of the superficial characteristics of terrain such as hills and valleys. Redistribution of earth materials (over millions of years) by the gradational forces helps to bring about further tectonic movements.

STUDY 7-1

In order to appreciate the time span of the different geological eras and periods, equate the history of the earth with a much shorter time period—a day. Let twenty-four hours represent the 4.5 billion years that the earth is thought to have existed. Using midnight to represent the beginning of earth history, determine the time when the first plant came into existence, when the Precambrian era ended, when the Appalachian mountains were formed, and so on. Selecting geological and biological events of importance, continue through the geological time chart down to the development of human beings (about 1 million years ago). For how long in the day have humans been in existence?

THE INTERNAL STRUCTURE OF THE EARTH

An understanding of the surface characteristics or landforms of the continents must begin with an examination of the origin of the tectonic forces that create the landforms. Because tectonic forces originate within the earth, we need to know something about its crust and interior before discussing the ways in which these forces affect the crust.

Some scientists think that the earth was originally a white-hot gaseous sphere which gradually cooled to a molten mass of metal and rock within a gaseous envelope. As the earth cooled further, some of the gaseous elements combined with other elements at the surface to form a solid but rather brittle outer *crust*, also known as the *lithosphere* (Fig. 7-2). Two other major zones—the *mantle* and the *core*—were formed beneath the crust when the original molten material solidified.

The nature of the earth's interior has always been puzzling and we have not yet penetrated more than a few thousand metres into its mass. Information about the interior has had to be obtained indirectly, because of the great temperatures and tremendous pressures beneath the surface. Most of our knowledge comes from data derived from the measurement and study of earthquake waves and man-made explosions. *Seismology*, which is the science of earthquakes, measures vibrations or shock waves as they travel around and through the earth. Changes in the speed of the shock waves and their paths through the earth can be interpreted to give information about conditions and structures within the earth. Much has been learned from seismology about the earth's interior, but a great deal still remains unknown or unproved.

Seismologic measurements have shown that the crust or lithosphere includes a continuous shell, the *sima*, and a discontinuous shell called the *sial*. The sima is a dense basaltic layer (silica and magnesium) about 5 to 8 km thick. It underlies the continents (sial) and forms the ocean floors. The sial is a lighter granitic layer (silica and aluminum) 20 to 40 km thick, forming the foundation of the continental masses. (The density of sial is $2700 \text{ kg}/\text{m}^3$ while the sima is $2900 \text{ kg}/\text{m}^3$.) The total thickness of the crust is estimated to be between 10 and 50 km; it forms less than 1 per cent of the earth's total mass. The crust is broken into many sections of varying size and shape (see Fig. 7-9). These sections are referred to as *plates*. There are six major plates and a larger number of smaller ones. These plates include both sial and sima and can be described as floating on the substratum of semi-liquid rock materials that make up the top of the mantle. Virtually all of the earth's tectonic activity occurs as a result of very slow movement along the margins between plates. Such movement is examined further in the sections on Sea-Floor Spreading and Plate Tectonics beginning on p. 171.

The centre or *core* of the earth, which makes up about 16 per cent of the earth's volume and 32 per cent of its mass, is generally thought to consist of an outer molten portion and a solid interior. The core is probably made up of the elements iron, nickel, and perhaps silicon. It has a temperature between 2200°C and 2750°C and its pressure is several million times that of the atmosphere at sea level.

Between the crust and the core is the zone known as the *mantle*. It has a depth of approximately 2400 km and constitutes 83 per cent of the earth's volume and 68 per cent of the earth's mass. Earth scientists believe that the particular combinations of temperature and pressure in this zone cause the mantle rock to be quite rigid but, nevertheless, in its upper part at least, capable of

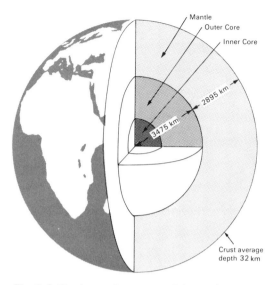

Fig. 7-2. The internal structure of the earth.

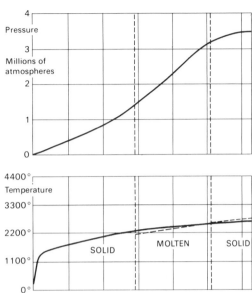

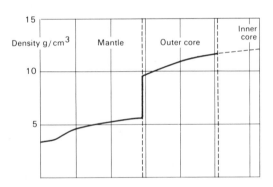

extremely slow movement. The mantle might be compared to the ice in glaciers, which will shatter when struck but under given conditions will flow very slowly. Separating the crust and the mantle is a zone called the Mohorovičić discontinuity (often shortened to 'M' discontinuity or *Moho*). It is named after the Yugoslav seismologist who in 1909 discovered that earthquake waves increase abruptly in velocity when passing through this layer.

The plates of the crust are in a sense floating on the mantle, somewhat like ice in water. Materials are continuously being added to or taken away from the surface of the plates, causing them to be in a state of extremely slow but constant vertical movement. In this way a state of equilibrium between the crust and the mantle, known as *isostasy*, is maintained. For example, we know that during the Ice Ages great ice sheets pushed the crust downwards, but when the ice disappeared the crust rebounded, or moved back upwards. The same kind

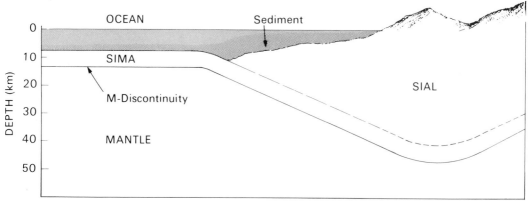

Fig. 7-3. Cross-section showing the division of the crust. The boundary between the sial and sima (the Conrad Discontinuity) is known only from seismic evidence because the greatest distance we have yet drilled into the crust is 8 km.

of crustal movement has occurred in areas undergoing erosion. The wearing away of rock materials in such areas has lightened the crust, causing it to lift. On the other hand, where the eroded rock materials were deposited, weight was added to the crust. This produced depressions or downwarping. Examples of this type of crustal movement can be found on many continental margins, such as the lower Mississippi River basin. Isostatic adjustments such as these, together with horizontal shifting of the plates (explained later in this chapter), subject the rigid crust to stresses and strains. These stresses and strains (tectonic forces) have warped and fractured the crust. Together with the gradational forces they are responsible for the enormous variety of terrain on the earth's surface.

THE COMPOSITION OF THE CRUST

The outer covering of the earth is composed of many different types of rocks; they in turn are composed of one or more minerals. For example, the rock limestone is made up mainly of the mineral calcium carbonate, although other common minerals such as quartz and feldspar are usually present. Minerals may be defined as naturally occurring inorganic solids. They are usually chemical compounds composed of two or more elements in combination. Minerals possess unvarying physical properties, notably a distinctive crystalline form, a particular structure, and a mode of fracturing. Although over 2000 minerals have been discovered, a mere ten of them make up over 90 per cent of the rocks of the crust. Rocks are simply an accumulation of minerals in a solid state.

All matter in the universe—which includes living things—is composed of 92 naturally occurring elements. These range in size from the lightest and simplest, hydrogen, to the heaviest, uranium. (There are also 12 artificially produced elements.) All elements are pure substances possessing certain physical and chemical properties. Of the 92 found on earth, eight make up just over 98 per cent of the crust (Fig. 7-4). Most elements combine chemically to form minerals although some —such as gold, copper, or sulphur—may exist in a pure or *native* form.

Fig. 7-4.

	Weight per cent
Oxygen O	46.6
Silicon Si	27.7
Aluminum Al	8.1
Iron Fe	5.0
Calcium Ca	3.6
Sodium Na	2.8
Potassium K	2.6
Magnesium Mg	2.1

Most minerals are formed from combinations of the elements oxygen and silicon (a non-metallic element of common occurrence in the compound silica, or silicon dioxide.) Quartz, one of the more common minerals in the crust and the most abundant material in sand, is mainly silicon dioxide (SiO_2). It is the hardest of the many rock-forming minerals and it is chemically stable as well. The aluminum silicates—feldspars, micas, clay minerals, etc.—are fairly common silicon compounds that are physically hard but subject to chemical decomposition. Other common rock-forming minerals include various oxides (notably iron oxide), chlorides, sulphates (notably gypsum), and carbonates, especially calcium carbonate and magnesium carbonate (dolomite).

Rocks

A general understanding of the rocks that make up the crust—and in particular the sial —is important to the study of landforms. In the following section a simple classification of rocks is presented. Rocks can be divided into three broad categories depending on their genesis or origin. The categories are igneous, sedimentary, and metamorphic.

IGNEOUS ROCKS

Igneous rocks are formed directly from the cooling and solidification of molten rock (*magma*). They are composed of minerals in the magma that have crystallized. (The name *igneous* is derived from the Latin, and means fiery.) There are two main types of igneous rocks. The first includes rocks formed by the solidification of molten material from the earth's interior that has found its way to the surface. These are known as *extrusive* or *volcanic* rocks. A familiar example of a molten material is lava. Most of us have seen pictures of it flowing from the crater of a volcano or from vents on its slopes. A number of different rocks are formed from the lava; some of the more common are basalt, andesite, rhyolite, obsidian, and pumice. Although all of these rocks can be produced by molten lava, they differ in mineral composition and texture, and such differences are reflected in their appearance. Pumice, for example, is usually porous, soft, and easily eroded, whereas basalt is much harder and more resistant to weathering. Large areas of the earth are underlain by volcanic rock. In some places these rocks were formed by outpourings of lava. For example, lava deposits cover thousands of square kilometres in the basin of the Columbia River of the northwestern United States, the western part of the Deccan peninsula of India, and parts of southern Brazil and Paraguay. Volcanic rocks can also be formed from the consolidation of the ashes produced by volcanic eruptions.

The second group of igneous rocks includes rocks that have cooled slowly underground, often at great depths. These are known as *intrusive* or *plutonic* rocks. They can be distinguished from the volcanic group by their coarser grain and larger crystals. These characteristics are the direct result of a slower rate of cooling. Granites (the most abundant rocks of this group), diorites, and gabbros are examples. Intrusive rocks are

Fig. 7-5. Massive sedimentary strata exposed along the Siletz River in Oregon.

very widespread. They occur in mountain systems, where erosion has exposed them, or in areas such as the Canadian Shield, where they exist on the surface as the remaining traces of mountains worn down during the Precambrian era hundreds of millions of years ago.

SEDIMENTARY ROCKS

For the most part sedimentary rocks are secondary or derived rocks. They are formed in a variety of ways from the deposition of material on the floors of lakes or the bottoms of seas. (Sedimentary rocks derive their name from the Latin *sedimentum*, meaning a sinking or settling down.) The principal source of the sediments is the erosion of pre-existing rocks. (Other methods of formation are described below.) Accumulations of rock fragments, sand, silt, clay, lime, or mud, are changed into solid rock by the mass of overlying sediments and then cemented together. The latter part of the process, known as *cementation*, is the more important stage in the conversion. It involves soluble 'cements' such as calcium carbonate, silica, or iron oxide. These cementing agents fill the pore spaces between the individual particles of sediment and bind them together.

Sedimentary rocks are built up by the slow deposition of material over extremely long periods of time. They are generally found in layers referred to as *strata* (Fig. 7-5). Strata may vary in thickness from thin sheets, known as *laminae*, to massive beds tens of metres thick. This variation is usually a result of discontinuous deposition, that is, the layers were formed at different periods. In addition, particles deposited in one period may have a different composition from those deposited in another period. Variations in grain size and colouring are the most common distinguishing features between strata.

There are three main categories of sedimentary rock. The first group, the *clastic* or *fragmental*, includes those rocks formed from non-soluble fragments of pre-existing rocks. These fragments, in the form of gravel, sand, silt, or clay, are deposited principally by water. During deposition they are roughly sorted according to size and mass. The larger particles, which are deposited first, are laid down at the edge of a lake or sea; the smaller particles are carried farther out before they settle. The largest of these fragments—sand or gravel over 2 mm in diameter—form the rock called conglomerate; grains of sand (0.6 mm to 2 mm) form sandstone; particles of silt (0.004 mm to 0.6 mm) form siltstone (mudstone); and particles of clay (less than 0.004 mm) form shale.

The second large group is *chemical precipitates*. It includes rocks formed from certain soluble materials that are deposited on the floor of a lake or sea either as the remains of shellfish, coral, and other marine organisms, or as precipitates left behind by the evaporation of water. Examples include salt (sodium chloride), gypsum (calcium sulphate), potash (potassium chloride), dolomite (calcium-magnesium carbonate), chert (a term covering a multitude of silicates), and borax (sodium pyroborate).

Organic rocks, formed from the remains of plants and animals, constitute the third major group of sedimentary rocks. An obvious example of this type of rock is coal, which is the remains of partially decomposed land plants. Coal is usually formed from vegetation that has decayed in swamps. As the decaying matter loses hydrogen it progresses through a series of stages from peat to lignite to bituminous coal, possibly anthracite, graphite, and finally pure carbon. Limestone is the most abundant rock of the organic group of sedimentary rocks. It is formed from the remains of tiny marine animals such as coral (whose shells are made of calcium carbonate), from other lime-secreting organisms, and from chemical precipitates. Such deposits accumulate under water where compression and, in particular, cementation slowly transform them to rock. Chalk is also an organic rock, formed from the remains of minute single-celled creatures (whose tiny shells are also made of calcium carbonate).

Sedimentary rocks are found over approximately 75 per cent of the surface of the earth, but they form only 5 per cent of the total volume of the crust. Shale, sandstone, and limestone make up approximately 99 per cent of all sedimentary rock. Some of the larger areas of sedimentary rock underlie plains such as those in the interior of North America and those in western Europe and extend into the Soviet Union.

METAMORPHIC ROCKS

Metamorphic rocks are rocks that were originally igneous or sedimentary but have been changed in composition, texture, and structure by great heat, pressure, or chemical activity within the earth. One cause of this change (metamorphism) is the extremely intense heat released from molten rock that has intruded into sedimentary or igneous rock. Metamorphic rocks may also be caused by the immense pressures that are exerted by tectonic forces on crustal rocks.

Metamorphism causes the minerals in sedimentary or igneous rock to recrystallize and rearrange themselves, thus forming a new type of rock. Very often metamorphic rocks can be distinguished by their *foliation* or banded structure, which appears as layers in the rock. Sometimes, where foliation is very pronounced (as in slate), planes of weakness are formed, along which the rock will readily split. Other metamorphic rocks, such as marble or quartzite, are *non-foliated* or homoge-

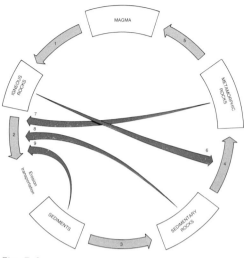

Fig. 7-6.

1 Figure 7-6 shows all the possible changes that might occur when one type of rock is transformed into another type. Each of the numbered arrows indicates a change. For example, Arrow 1 represents the cooling of molten rock to form igneous rock. What changes do the other arrows represent? Relate the changes shown in the diagram to those in Fig. 7-7 and explain the processes that are involved in each transformation. List examples of rocks that may be formed in each of the changes.

2 Why do sedimentary rocks make up such a high proportion of the surface of the crust (75 per cent) and such a low proportion of its volume (5 per cent)?

CONTINENTAL DRIFT

As knowledge of the earth's structure has increased, geophysicists have developed new theories to explain the characteristics of its surface. These theories range from accounts of how the earth's hills and valleys were formed to explanations of the shape and position of the continents themselves.

For centuries people have speculated about the apparent correspondence between the coastal outlines of the continents, particularly those of the east coast of the Americas and the west coast of Africa and Europe. A world map will show, for example, that the great bulge on Brazil's coast seems to fit rather well into Africa's Gulf of Guinea. Early in this century a German meteorologist, Alfred Wegener (1880-1930), theorized that the present continents were all once part of one super-continent that he called Pangaea. According to Wegener's theory, this land mass began to split up about 200 million years ago as a result of internal pressures in the earth. The various parts of Pangaea, including the present continents, slowly drifted to the positions they occupy today.

neous. There are more varieties of rocks in the metamorphic group than in either of the other groups. Almost every igneous and sedimentary rock has a metamorphic equivalent. For instance, granites and diorites may change to gneiss (pronounced 'nice'), sandstone to quartzite, and limestone and dolomite to marble. The changes, however, are seldom direct transformations, as these examples may suggest. There are different grades of metamorphism depending on the degree of temperature and pressure. For example under low temperatures and pressure shale changes to slate, but under high temperatures and pressure shale changes to schist, and later to gneiss.

Metamorphic rocks are found in many locations but occur particularly in mountainous regions and in large shield areas such as the Canadian or Baltic Shields. In these areas they are thought to be the result of widespread and complex tectonic activity during the Precambrian era.

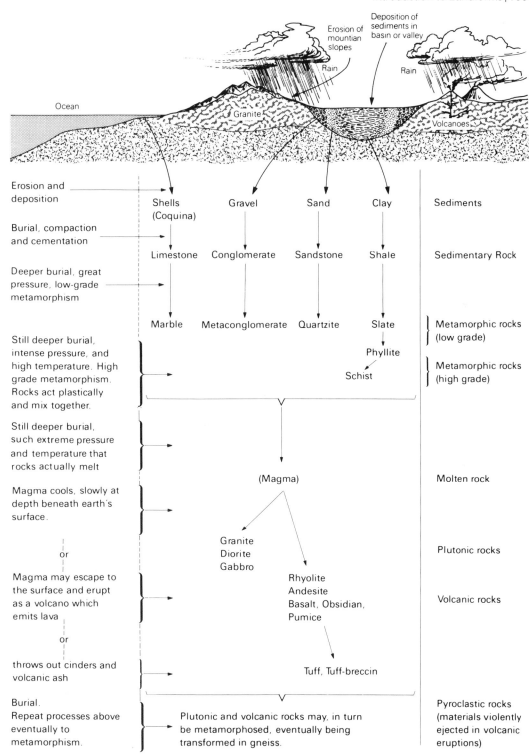

Fig. 7-7. The formation of rocks (from Putnam and Bassett).

Wegener based his theory on (1) the similarities in geological structures between continents that are now separated by a large expanse of ocean, and (2) the similarity of fossil remains of ancient forms of life found on the separate continents. The latter observations were not new however. Biologists and other scientists before Wegener had explained the similarity of such fossil remains by a theory of land bridges. They suggested that at one time land bridges must have connected the continents, across which animals migrated and plants spread. This idea of land bridges was based on a widely held belief that the earth was slowly cooling and at the same time shrinking: as the earth contracted, the oceans increasingly encroached on the land and gradually submerged the land bridges. The contraction theory explained not only land bridges but also irregularities such as mountains and hills, which were supposedly formed like wrinkles on the contracting surface. Wegener recognized, however, that the materials forming the ocean floor (sima) were different from those forming the continents (sial). Furthermore, he also noted that the density of the sial was approximately 90 per cent that of the sima. Such evidence suggested to Wegener that land bridges never existed, and that because of their lower density the continental masses had reached their present position by floating like enormous ice floes on a solid sea of sima.

Wegener's theory was much discussed, ridiculed, and subsequently dismissed by most scientists. How could it be possible for continents to drift through a solid and rigid mass of sima? What forces were responsible for such movements, and what would happen to the sima ahead of and behind the moving continents? Unfortunately Wegener was not able to supply satisfactory answers to these questions. It was not until the 1950s that a series of scientific discoveries provided much of the evidence Wegener had lacked. Some of these discoveries are outlined below.

1 Sensitive instruments for measuring earth magnetism were developed. These indicated that millions of years ago, when molten iron oxides in the crust cooled, they were fixed pointing towards the magnetic north pole. The surprising discovery, however, was that in North America these magnetic 'needles' pointed to a different magnetic-pole location than the needles in rocks of the same age in Europe. This would suggest that either the earth once had two separate magnetic poles (a highly unlikely possibility), or the position of the two continents had changed. In fact, it was found that if the continents on both sides of the North Atlantic had been joined together when these oxides were fixed in position millions of years ago, the magnetic needles on both continents would then have pointed towards a single magnetic pole.

2 A major program of mapping the ocean floor was carried out by the United States in the 1950s. This resulted in the accurate mapping of a chain of ridges and mountains, 75 000 km long, under the world's oceans. The ridges, 1500 m to 3000 m higher than the adjacent ocean floor, are cut by faults (Fig. 7-8) and divided into numerous short, straight sections that gradually change direction in a step-like fashion. In the Atlantic the ridge is almost exactly midway between the Americas on the west and Europe and Africa on the east. Geophysicists could no longer pretend that it was merely coincidence that the ridges were parallel to the coastlines. The location of the ridges had to be related to the spreading of the continents.

3 The underwater mapping program also provided the first accurate outline of the continental shelves (that part of the continent covered by the ocean and seas). This outline showed that the shelves of adjacent continents fit together much closer than anyone had imagined and much better than the shoreline fit on which Wegener had based his theory.

4 Seismological records were also improved

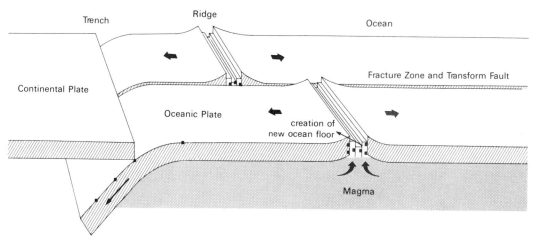

Fig. 7-8. The diagram illustrates the rift or ridge zones to the right and the subduction zone to the left. Note that the rift is bordered by ridges pushed up by the upwelling magma. The trench or subduction zone is formed whan an oceanic plate meets a continental plate. Transform faults occur when two plates are moving past each other. The dots on the diagram represent the earthquakes that accompany these tectonic processes.

markedly in the 1950s and 1960s. This enabled scientists to locate earthquakes more exactly. The records showed that earthquakes are concentrated in a much smaller area than had formerly been suspected. They occur mainly along the deep ocean trenches, the ocean ridges, and at other points of contact between the plates of the earth's crust (Fig. 7-9). This suggested that crustal movement was taking place in such areas.

Sea-Floor Spreading

As a result of the new evidence that had emerged over the past few years, the old theory of continental drift has not only been revived but virtually substantiated. Recent seismographic discoveries have revealed a layer of low-density rocks in the upper mantle. This layer is known as the *asthenosphere*. It is a relatively soft layer 95 to 240 km thick.

The asthenosphere is capable of very slow movement because its temperatures are higher and its pressures are lower than those in the mantle beneath it. The asthenosphere is considered to be the transporting medium for continental drift. The generally accepted explanation of the movement in this zone is that heat from the earth's interior causes convection currents in the asthenosphere, which in turn produce a dragging force on the bottom of the crust that causes it to move slowly. As the currents approach the surface, and as pressure decreases, pockets of molten rock are formed. Some of this molten rock is extruded from the mid-ocean rifts, where it solidifies and actually becomes part of the ocean floor (sima). The crust itself moves away from the rift areas at rates ranging from 1 to 7 cm per year. While these amounts may seem small, only 6 cm of movement per year (3 cm on each side of the rift) would have

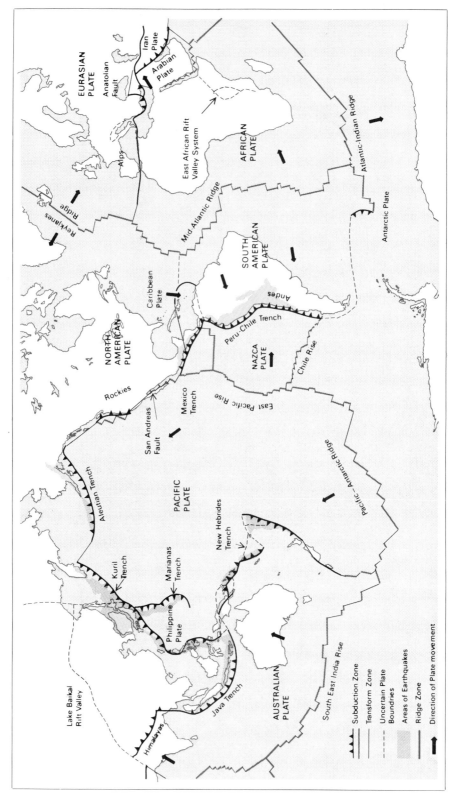

Fig. 7-9. Map showing the direction of the earth's crust into major plates.

been sufficient to form the entire floor of the Pacific Ocean in a hundred million years. This is a relatively short time in terms of total earth history.

It is reasonable to suppose that if new ocean floor is being created along the mid-ocean ridges, then an equal amount of crust is being destroyed somewhere else. Figure 7-8 shows how this destruction occurs. The earth's crust is made up of six major sections, or plates, and several minor ones. Unlike the asthenosphere beneath, the plates are cold, quite rigid, and capable of sliding over the hot and partially molten asthenosphere. The arrows on Fig. 7-9 show that along their boundaries some of the plates approach each other head on. When this occurs the plates either slide laterally past one another or one plate remains horizontal while the other is bent downwards at a fairly steep angle. In the latter case the depressions formed by the bending plate are the deep-ocean trenches. The bottom plate is pushed very slowly into the asthenosphere and the movement of rock against rock produces earthquakes some distance below the earth's surface. The rate of crustal destruction clearly must balance the rate at which new crust is formed at the mid-ocean ridges. Also associated with the collision of plates and the oversliding process are the major mountain systems of the world (discussed in the next section). In addition, chains of volcanic islands are also formed adjacent to the trenches. These are particularly common in the western Pacific. The formation of such volcanic islands is thought to begin when less dense materials in the lower plate melt (as the plate is forced into the asthenosphere) and such materials are then forced upwards through cracks in the oversliding layer to form volcanoes. The islands are completely formed when the volcanic material projects above the surface of the ocean. Similar volcanic action along the mid-ocean ridges can also form islands, although the number of islands

formed in this way is smaller. One example is the island Surtsey, off the coast of Iceland (which is itself a volcanic island). It emerged in 1963 as a result of volcanic eruptions from the mid-Atlantic ridge 100 m below the surface of the sea.

The theory briefly outlined above was put forward in 1960 by an American geologist, Harry Hess. While it received wide acceptance, six years passed before additional evidence was uncovered. The evidence related to the discovery that the earth's magnetic field reversed itself at irregular intervals every few hundred years. The reversals, it was argued, should be apparent in the magnetic orientation of the lava extruded from the ridges. Studies by geomagnetists confirmed this contention by showing that alternate strips of positive and negative rock magnetization existed on each side of the ridge. This proved that sea-floor spreading had been continuous throughout the Cenozoic era and had occurred at a rate of about 3 cm per ridge flank per year (a total of 6 cm for both sides of the rift). This evidence has convinced most scientists that the theory of global or plate tectonics is indeed a plausible one. However they will admit that it provides only the basis for a total expansion of continental drift. There is certainly still a great deal that is not completely understood.

Plate Tectonics

The term 'tectonics' refers to the origin of the structural features of the earth's crust. Plate tectonics, therefore, is the explanation of these features in relation to the movement of the earth's plates. As already described, the surface of the globe is made up of a mosaic of plates of varying sizes. Most of these plates include both continental crust (sial) and oceanic crust (sima). The complex movement of the plates produces three different types of zones or boundaries where plates meet: a *subduction zone*, where one plate slides below another into the mantle and is consumed by it; a *ridge or spreading zone*, where new oceanic crust is being created; and a *fault zone*, where the plates slip laterally past one another.

As explained in the previous section, the plates are moving because new ocean crust is being created at the ridges while old ocean crust is being destroyed at the trenches. Though most of the plates contain both continental and oceanic crust, it is thought that only the oceanic crust is involved in the creation-destruction process. This theory is substantiated by the fact that the oldest known rocks on the continents are approximately 2.5 billion years old, while the oldest known rocks on the ocean floor are less than 200 million years. Though the oceanic crust is easily consumed in the ocean trenches, it appears that when the less dense continental areas reach the trenches they resist consumption, and subduction is terminated.

All plate movements, and particularly that which occurs in the subduction zone, give rise to tectonic forces. Of special consequence are the forces generated when two continental masses collide. Since the continents cannot be subducted, the collision causes the rock between the plates to be compressed and ultimately to buckle or fold into a mountain range. Most of the world's major mountain systems probably originated in this way.

The Atlantic Ocean can be used as an illustration. Figure 7-10 shows various stages in the development of the Atlantic over the past 600 million years.[2] In (a), the late Precambrian era, a rift developed and North America and Africa began to split apart. Through the process of sea-floor spreading an ocean opened up where the Atlantic Ocean now exists and sediments were deposited on the continental shelf and the ocean floor adjacent to each continent. The sediments slowly built up to considerable thicknesses (up to 16 km) forming a structure known as a *geo-*

syncline (b). The weight of these sediments in the geosyncline caused the crust to be depressed isostatically and the sediments remained below the surface of the sea. In (c) the ancestral Atlantic Ocean began to close as a result of several factors. Among these, the weight of sediments caused the lithosphere to break, creating a subduction zone adjacent to the North American continent. As a result of the subduction process, sediments in the geosyncline were compressed and folded. The entire mass was thrust upwards, thus adding new materials along the continental margins in the form of huge mountains. These processes were accompanied by the shattering or faulting of the sediments and the intrusion of molten rock. Eventually the continental masses collided and were joined together (d), further compression and folding occurred, and the subduction process ended. It is estimated that southeastern North America and Africa were joined by such a process between 225 million and 350 million years ago. The Appalachian Mountains were formed and their area added to the North American continent. About 180 million years ago (e), the present Atlantic Ocean began to open along the old ridge line, thus starting the creation of new geosynclines.

Another example of the process of plate tectonics is the collision of India, a part of the Australian plate, with the southern margin of the Eurasian plate (see Fig. 7-8). The sedimentary materials that had formed in the geosyncline between these two plates were compressed and folded into the landforms we now know as the Himalayan and associated mountain ranges.

STUDY 7-3

1 Examine the figure showing major crustal plates and their direction of movement (Fig. 7-9) and a world map in an atlas. Comment on the relationship between the leading edge of these plates and major mountain systems.

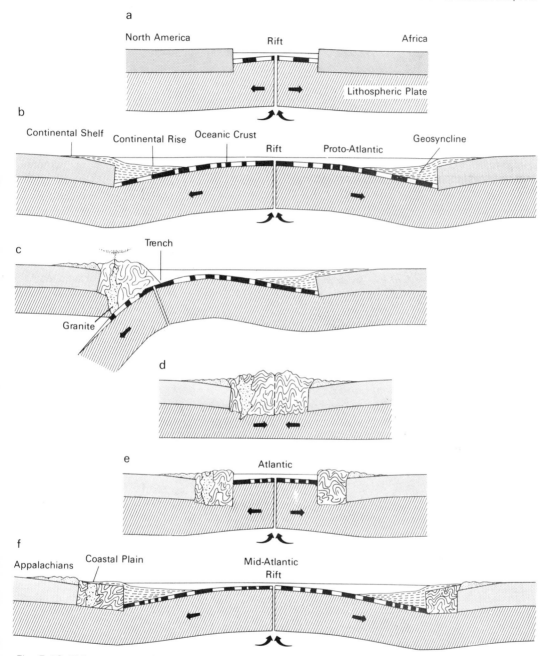

Fig. 7-10. This sequence of cross-section diagrams shows the history of the North Atlantic from the Pre-Cambrian to the present and the effect of the opening and closing of the ocean basin on the formation of mountains. In (a) and (b) the process of sea-floor spreading is opening the ancestral Atlantic with huge sedimentary deposits (geosynclines) forming on the margin of each continent. With (c) the ocean begins to close, the lithosphere breaks, and a trench is formed. The ocean plate descending into the mantle causes the sedimentary basin on the west side to be folded and fractured, producing what are now the Appalach-ian Mountains. By (d) the continents have come together again, causing the sediments on the African side to be folded also. In (e) (about 180 million years ago) the ocean begins to reopen and is still doing so at the rate of 3 cm a year. At the same time new sedimentary basins are forming off both coasts (f).

2 The Red Sea has been described as an incipient ocean. What does that mean in terms of plate tectonics?

3 Why is there so little continental shelf off the west coast of the Americas as compared to the east coast?

4 If we could look ahead 100 million years, how might the Atlantic Ocean and the eastern seabord of North America differ from their present form?

FAULTING, FOLDING, AND VULCANISM

The tectonic forces, which have just been explained in relation to the movement of plates, influence the surface of the earth through the processes of *faulting, folding,* and *vulcanism.*

[Folding and faulting include] those processes which are involved in the breaking, bending, and warping of the earth's crust and the elevation, depression, or displacement of one part with respect to another. Vulcanism includes those processes which involve the transfer of molten material from one place to another within the earth's crust or its expulsion at the earth's surface. The ultimate tendency of the tectonic forces and their processes is to cause differences in surface elevation on the earth, and by heaving up the crust here and depressing it there or by pouring out upon it great masses of molten lava, to construct surface features of great height and areal extent.[3]

Though tectonic activity is as old as the earth itself, the present major mountain systems are mainly a result of mountain building during the Cenozoic era. Vestiges of earlier periods of tectonic activity are evident in some of the earth's lesser systems of mountains and hills.

Faulting or Crustal Fracture

Throughout the entire history of the earth the various stresses and pressures associated with plate tectonics have produced numerous fractures in the earth's crust. Most of them are very small and stop a short distance beneath the surface. In some places, however, continuous pressure has caused deep cracks to form. Eventually this pressure will cause the movement or displacement of rocks along the plane of the fracture. Such cracks are called *faults.* The movements causing faults probably occur very suddenly, but the displacement that results from such movements at any one time may be only a matter of a few centimetres. (Exceptional horizontal displacements as large as six metres were observed along the San Andreas Fault after the 1906 San Francisco earthquake, but such displacements—whether vertical or horizontal—are not considered typical.)

There are many different kinds of faults. Transform faults (p. 174) are one of the more important types. These occur along the plate boundaries where one plate or part of a plate is sliding laterally past another. Because the movement is essentially a horizontal one, transform faulting seldom creates major landforms. Block faulting, on the other hand, does produce significant landforms. It is caused by vertical movements of large sections of the crust along faults or planes of fracture. These movements, in turn, are caused by the forces generated by moving plates as well as isostatic adjustment. In simple terms, some parts of the crust may rise and other parts may settle. The landforms that result from such block faulting vary from the simple ones shown in Fig. 7-11 to more complex types. Depressions sometimes known as rift valleys are an example of the latter. These are formed when a part of the crust settles between parallel faults. The Owens Valley shown in Fig. 7-12 is an example of a rift valley. Other prominent examples include the Rocky Mountain Trench, the East African Rift Valley, and part of the Rhine Val-

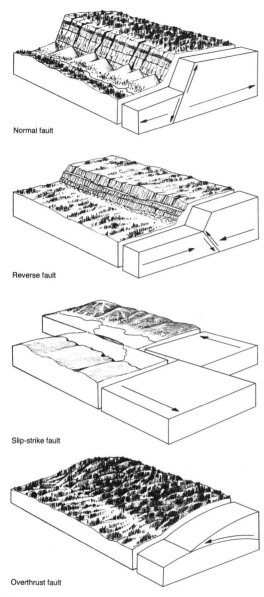

Normal fault

Reverse fault

Slip-strike fault

Overthrust fault

Fig. 7-11. Four types of faults. The right-hand portion of each diagram shows how, in theory, the faulting occurred; to the left of this is shown the landscape as it might appear at present.

ley in West Germany. Faulting that is even more complex than rift valleys occurs when some parts of the crust are raised or tilted and others are depressed. The Basin and Range area of eastern California and Nevada was formed in this way. The western part of this area, the Sierra Nevada mountains, is shown in Fig. 7-12. These mountains were formed when a huge section of crust was forced up along the faults at the eastern boundary of the Sierra Nevadas. Thus, these mountains have a very steep east-facing slope and a much more gentle west-facing slope.

Since the formation of any substantial landform requires long periods of faulting, the present surface characteristics of such landforms are mainly a result of gradational action, and the original faults are often quite difficult to identify.

Folding or Crustal Bending

The compression or folding of the lighter sedimentary rocks of the crust is related to the movement of the earth's plates, though folding occurs under somewhat different circumstances from faulting. The size of the folds and the area of land involved vary greatly. In some instances the rocks have been compressed into complicated folds of mountainous proportions. In other instances, large areas have been slightly warped—mainly through isostatic adjustment—so that some parts of the crust have been raised and others lowered. Such crustal movements have been accurately surveyed in some areas, revealing rates of uplift and depression ranging from a fraction of a millimetre to as much as ten millimetres per year.

Severe folding has been responsible for the formation of most of the world's mountain ranges. As we have already seen, folding originated when extensive deposits of sediment accumulated in large depressions in the earth's crust (geosynclines). The deposits

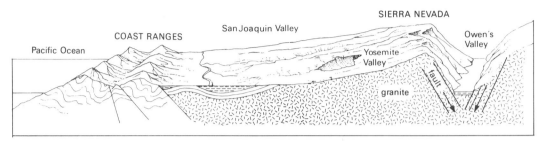

Fig. 7-12. Mountains such as those of the Sierra Nevada (known as fault-block mountains) are the result of the elevation of large blocks along marginal faults. The Owens Valley is a *graben* or rift valley formed through down faulting. Such valleys are usually trench-like, with fairly steep walls (fault scarps) rising to adjacent highlands.

Fig. 7-13. The San Adreas Fault in California is shown running through the middle of this photograph. The stream channel to the left is offset about 150 m in a right lateral movement along the fault.

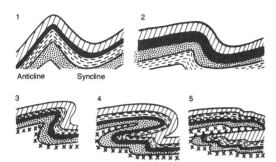

Fig. 7-14. Types of folds: 1 simple; 2 asymmetrical; 3 overfold; 4 recumbent anticline; 5 overthrust or nappe.

metrical to very complex, recumbent ones (Fig. 7-14). (Faulting and vulcanism are often associated with the latter type of fold and greatly complicate its structure.)

Folding, like faulting, takes a long time to produce major landforms. However, while such landforms are slowly taking shape they are also being worn down by the gradational forces. The cross section in Fig. 7-15 illustrates different types of folds and the surface profile produced by gradation.

Severe and spectacular folding has gradually created mountain ranges but moderate folding—called *warping*—has influenced much larger areas of the earth's surface. Though warping is often initiated by plate movement, subsequent erosion and deposition on the warped surface may further intensify the warping by triggering an isostatic adjustment. Owing to the large areas involved and the very slight bending that occurs, warping seldom produces any prominent surface irregularities. One common result of warping is the slow emergence or submergence of coastal areas such as the lands around the North Sea. Elevation measurements taken over several centuries show a signficant downward bending of the land

were subjected to intense compression as a result of plate movement—particularly the collision of major plates (see page 173). The amount and type of folding vary from one mountain system to another and even within the same range. Such variations may be attributed to a complex series of events related to the amount of compression. Several different types of folds are recognized, ranging from those that are simple and sym-

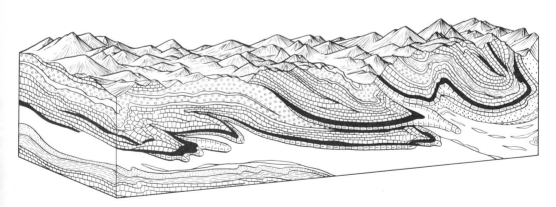

Fig. 7-15. The Alps are among the most complex mountain structures known. They include a mixture of simple anticlines and synclines, recumbent anticlines and nappes, as well as faults — particularly of the thrust variety. The rocks making up these mountains are similarly complex and include igneous, sedimentary, and metamorphic types, ranging in age from Precambrian to Tertiary.

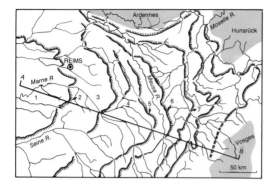

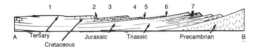

Fig. 7-16. The scarplands of the Paris Basin are here shown in plan and cross-section views. Some of the more prominent scarps and adjoining plains are numbered on both the map and the diagram. 1 Ile de France; 2 Falaise de l'Ile de France; 3 Champagne Pouilleuse; 4 Champagne Humide; 5 Argonne; 6 Côtes de Meuse; 7 Côtes de Moselle. (Coloured areas indicate higher elevation.)

surface in parts of Great Britain and the adjacent areas on the continent.

Another example of warping can be seen in Fig. 7-16, which shows the gently tilted sedimentary strata of the Paris Basin in France. As illustrated in the diagram, the gradation wore away the upper part of the warped surface, leaving a series of eastward-facing scarps, which in some places are prominent cliffs and in other areas only steep slopes. The scarps have remained because a cap of resistant sedimentary rocks protects less resistant sedimentary rocks beneath it.

STUDY 7-4

1 Figure 7-17 illustrates the fairly simple and symmetrical folding that has occurred in the Appalachian region of the United States.

a Find the location of the cross section on the photograph. (An explanation of stereo photographs is found on p. 296).

Fig. 7-17. The air photograph and cross section (A-B) show a part of the Loysville, Pa. topographic map. This landscape is fairly typical of that part of the Appalachians known as 'the ridge and valley province'. Photo scale 1:56 000 (approx.).

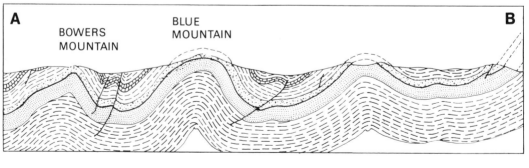

b Comment on the relationship between rock type and the landform structure in Fig. 7-17. Remember that sandstone and quartzite are more resistant to erosion than limestone and shale.

c Make a second cross section parallel to and west of the area shown in the first cross section.

2 a Make a west-to-east cross section sketch of the Paris Basin (Fig. 7-16), showing with dotted lines the position that the sedimentary strata would have taken if gradation had not occurred. Referring to an atlas, add a scale showing the approximate elevation.

b Referring to an atlas, describe the relationship between the landscape and the river system of the area. Why do the rivers cut through the scarps?

c Referring to the time chart on pp. 160, describe the geological history of the region.

Vulcanism

In addition to folding and faulting, the welling up of molten rock on or near the earth's surface also affects the outer part of the crust. The source of the molten rock in the asthenosphere has been discussed earlier in this chapter; earth scientists are still a long way from a complete understanding of the origins of molten rock.

Vulcanism is of two kinds (the distinction between them has already been noted in our description of igneous rocks, which are a product of vulcanism). *Extrusive vulcanism* is the eruption of molten rock on the earth's surface, especially from volcanoes. A great deal of molten rock, however, never reaches the surface but penetrates and solidifies in underground cracks and crevices. This is called *intrusive vulcanism*. Evidence of intrusive vulcanism can be seen on the earth's surface if the intrusion warps or bends the surface rocks, or if the erosion has stripped away the overlying rock. (Molten rock that reaches the surface is called *lava*, while molten rock under the surface is known as *magma*.)

EXTRUSIVE VULCANISM

Not many years ago it was believed that the earth consisted of a molten interior with a thin covering crust. Volcanoes were regarded simply as weak points in the crust where molten rock escaped. We know now, however, that most of the earth's interior is solid and that the sources of the molten rock are usually small shallow pockets within the crust. What, then, produces an eruption? Most earth scientists now believe that the solid rock of the mantle, which is heated from below, wells up in huge convection cells. These cells contain the heat energy and the raw materials that produce volcanoes as well as the various types of intrusive vulcanism that are described later. As this hot rock rises in the mantle the pressure on it is reduced. The temperatures at which solids melt depend on the pressures imposed on them —the lower the pressures the lower the melting points. For example, when the rock of the mantle rises from 300 to 80 km it will lose little heat, but the pressure on it will be greatly reduced. At this point the temperature of the rock may now be higher than the melting points of some of its constituent minerals. Partial melting will result, and as the proportion of liquid increases it will tend to segregate and perhaps separate from the minerals that do not melt. This liquid or magma is the parent material from which all volcanic products are derived.

The composition of the magma depends mainly on the depth at which it separated from the minerals that did not melt. However, many processes may alter its composition between the time of its formation and its eruption on the surface as lava. For example, the rising of the magma towards the surface

Fig. 7-18. Steeply tilted strata similar to those shown in Figure 7-15 can be seen in these mountains in Switzerland.

may be interrupted and the magma may be held at a particular level for a considerable time. During this interruption heat may be lost and some elements in the magma will start to crystallize, leaving the remaining magma very different in composition from the original. In addition, the rising magma may also partially or completely melt the crustal rock with which it comes in contact, adding these melted minerals to the magma. Different kinds of magma may be produced by either of these processes, or by a combination of both of them. This explains the many different kinds of lava erupted by different volcanoes or, indeed, by the same volcano at different times. The extrusion of lava on the earth's surface results in the formation of volcanoes or lava fields. Extrusive vulcanism occurs mainly in areas closely related to the distribution of earthquakes (Fig. 7-9), which are in turn related to the ridges and trenches at the margins of the plates. Of the 800 active volcanoes in the world (usually defined as those that have erupted in the last 2000 years), more than 60 per cent are located near the plate boundaries that rim the Pacific Ocean. Smaller concentrations are found along the ocean ridges. Volcanic activity, then, is closely related to regions where the crust is being either created or destroyed.

The eruptions that produce extrusive landforms vary greatly. The variations can be explained largely in terms of differences in the chemical composition of the molten material discussed above. Some lava is high in silica (such as andesite) and quite viscous. It tends to solidify when approaching the surface or after making contact with the atmosphere. Such lava is commonly accompanied by steam and various gases that cannot easily escape through the viscous lava. Pressure then builds up and a series of explosive eruptions takes place. Other volcanoes extrude lava that is lower in silica than andesite (for example basalt). After basalt reaches the surface, it may flow some distance before it solidifies. Gases escape from it more easily and thus explosions seldom occur.

There are many variations between the two extremes just described. Four major types of eruptions—explosive, intermediate, quiet, and fissure—and the landforms associated with each are described below.

EXPLOSIVE ERUPTIONS

On Monday, 27 August 1883, an island known as Krakatoa (Fig. 7-19), located in the Sunda Strait between Java and Sumatra, erupted several times so violently that the shocks were recorded all over the world. The noise of the explosion, first thought to be gunfire, was heard as far away as 4000 km at Mauritius in the Indian Ocean. Volcanic ash was hurled into the upper atmosphere and covered thousands of square kilometres of surrounding land and water. Giant sea waves called *tsunamis* (see p. 36) were set in motion. Their crests—which were over 30 m high—struck the coasts of Java and Sumatra, and at least 36 000 deaths from drowning were recorded. Today all that remains of Krakatoa's base is a broken ring of islands around an enormous cauldron-like cavity (a *caldera*) 300 m deep. Much of the rock that originally formed the volcanic cone

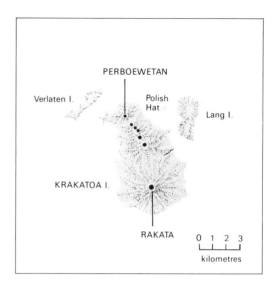

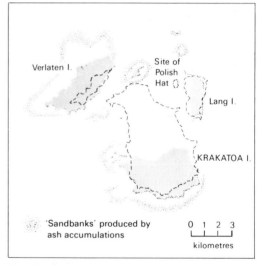

Fig. 7-19. Krakatoa and surrounding islands before (A) and after (B) the 1883 eruption.

was blasted into the atmosphere; the larger fragments fell back into the sea around the island. The rock of the remaining cone collapsed into the volcanic chamber that had been filled with molten rock before the eruption.

Another kind of volcanic explosion occurred in 1902 with the eruption of Mount Pelée on the Caribbean island of Martinique. The top of the 2250 m high mountain was blown off. Of greater consequence, however, was the dense, rapidly moving cloud of ash, dust, rock fragments, and hot gases (600-700°C), known as *nuée ardente* (glowing cloud). It moved down the slopes and killed all but two of the 30 000 people in the town of St. Pierre.

While these are extreme examples, they do illustrate some of the phenomena associated with explosive eruptions—tsunamis, the dust blanket, *nuée ardente*, calderas, and far-ranging earth tremors. Most of the existing explosive volcanoes, such as Lassen Peak in California or Mt. Katmai in Alaska, have undergone less severe explosions. Such volcanoes are normally steep sided and have a symmetrical cone formed from layer after layer of volcanic ash, cinders, and other material.

A small explosive eruption sometimes produces cinder cones. These cones are made of fragments of solidified lava—ranging from ash to large rock fragments—that exploded from a central vent. Such cones may grow very quickly but seldom reach more than 400 m. For example, Mt. Paricutín in Mexico reached a height of 325 m in the first year of its existence. Most cinder cones are small and usually occur on the sides and in the craters of volcanoes. They seldom erupt more than once.

INTERMEDIATE ERUPTIONS

Intermediate eruptions produce a great variety of different volcanoes—some built by the outpouring of lava, some by explosive discharge, and some by both. Usually these volcanoes are less steep than those developed entirely from explosions, and they vary greatly both in size and shape. Volcanoes created by both lava and explosive discharge are known as *strato* or *composite* volcanoes and can commonly be recognized by their graceful concave slopes (see Fig. 7-20). Mt. Vesuvius (1200 m) is an example. Its eruptions have included both the upwelling of lava and the explosive discharge of great quantities of such materials as steam, ash, cinders, and rock projectiles (called *volcanic bombs*). Other examples are Fujiyama (3700 m) in Japan, Etna (3300 m) and Stromboli (1926 m) in Italy, Popocatepetl (5452 m) in Mexico, and Mt. Rainier (4320 m) and Mt. Hood (3370 m) in the western United States (Fig. 7-20).

QUIET ERUPTIONS

Another kind of volcano, known as a *shield volcano* or lava dome, is formed by the extrusion of slowly cooling lavas. Such volcanoes have a central crater, but most of the lava is extruded from fissures on their slopes. The central crater, or caldera, apparently resulted in most instances from subsidence or settling that occurred as molten lava poured from reservoirs beneath the surface. Because of its basic composition, the lava from such volcanoes spreads outward from its source and may cover a considerable distance before finally solidifying. Even when nearing its solidification point the lava and gases can escape quite easily. Consequently, the lava pours out of the volcano intermittently over a period of days or even months rather than erupting in one or several large explosions. This is known as a quiet eruption. In certain cases these escaping gases spray liquid lava tens of metres in the air, forming plumes of incandescent liquid rock known as *fire fountains*. Volcanoes formed as a result of these lava

Fig. 7-20. The southeast face of Mt. Hood in Oregon. The three volcanic peaks in the distance are, from the left, Mt. St Helens, Mt. Rainier, and Mt. Adams. These are members of a chain of volcanoes in the Cascade Mountains extending across Washington and Oregon and into California.

flows have a very large base area, gentle slopes, and a rounded profile (Fig. 7-21). The Hawaiian Islands, which form part of a volcanic chain that rises from the floor of the Pacific, owe their existence to this form of vulcanism. For example, Mauna Loa on Hawaii rises 4100 m above sea level, extends at least 4500 m from the ocean floor to sea level, and has a circumference of about 320 km at its base below sea level.

FISSURE ERUPTIONS

Of all the earth-forming or tectonic processes fissure eruptions are undoubtedly the most important volcanic events. They occur along undersea ridges and have been responsible for almost all the crust beneath the oceans. The lava extruded from such eruptions is similar to that produced by quiet volcanoes such as Mauna Loa. It is, however, extruded from a great many fissures so that it covers the surface fairly evenly, forming a *lava plateau* or in some cases a *lava plain*. Iceland is an example of a lava plateau built up above sea level by fissure eruptions from the mid-Atlantic Ridge. As the sea floor spreads beneath the island, Iceland is being slowly stretched. The newly extruded volcanic material that solidified in tensional faults in the middle of the island causes this expansion (Fig. 7-9). The Columbia Plateau in the northwestern United States (Fig. 7-22) was formed in the same way about twenty million

Fig. 7-21. The destroyed town of Kapaho on the island of Hawaii. The Nakamura store is all that was spared by the spectacular 1960 Puma volcanic eruption. Before the eruption, the little Hawaiian village nestled in a lush green setting of sugar cane fields, palm trees, and papaya orchards.

years ago. Over a long period, a series of eruptions—averaging 10 m in thickness—occurred from thousands of fissures. These completely buried the original surface, filling valleys, covering hills, and ultimately forming a fairly uniform plain 130 000 km² in area and more than a kilometre in depth between the Cascade Range and the Rocky Mountains. Other landscapes formed from fissure eruptions are found in the northwestern Deccan Plateau of India, in the Paraná uplands of Brazil, and in Ethiopia and other parts of Africa.

The destructive effects of extrusive vulcanism are well known, but it must be pointed out that vulcanism has also been beneficial to humans. Soil derived from volcanic rock is usually rich in mineral nutrients, particularly potassium. Highly productive agricultural areas have been developed on volcanic soils in such places as Java, Hawaii, Sicily, and the Columbia Plateau of the United States.

INTRUSIVE VULCANISM

Much of the molten rock that moved towards the outer part of the crust slowly cooled beneath the surface, producing rock forms of many different sizes and shapes. The largest of these are called *batholiths*. They are enormous masses of igneous rock—sometimes hundreds of kilometres across—that appear to be of great depth. They are often found exposed in the centre of mountain ranges

Fig. 7-22. Erosion by the Palouse River has exposed the different basalt layers of a part of the Columbia Plateau. The steep walls and columnar jointing are characteristic of lava flows.

after the surface rock has been worn away. In southwest England and in Brittany such masses now stand out as upland areas. The original folds of the Permian period have been laid bare, leaving the harder igneous rock covered only by the thin soil and vegetation of moorlands. These moorlands drop sharply to the surrounding areas underlain by less resistant sedimentary rock.

A *laccolith*, which is a smaller, dome-shaped body of igneous rock, accounts for the landscape shown in Fig. 7-23. The intrusion of magma caused the outer sedimentary layers to warp and to form a dome. Erosion has since taken away the top of the dome,

exposing the laccolith and leaving in-facing ridges called hogbacks to mark the more resistant sedimentary layers.

Smaller intrusions such as *dikes, sills,* and *volcanic plugs,* as well as the larger laccoliths and batholiths, are illustrated in Fig. 7-24.

Earthquakes

An earthquake is a form of energy that causes wave motion to be transmitted through the ground surface. These shock waves range from weak tremors that are barely noticeable to severe ones capable of destroying buildings and opening fissures in the ground. In

Fig. 7-23. The Black Hills of South Dakota.

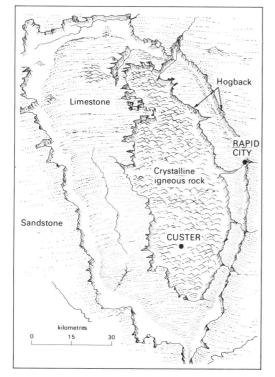

the most extreme cases the waves can actually be seen on the ground causing objects to be thrown in the air.

An examination of Fig. 7-9 will reveal that most earthquakes are closely related to the edges of the earth's plates and particularly to the leading edges, where one plate is sliding below another or two plates are moving laterally past each other. This suggests that the shock waves felt at the surface are a result of the sudden movement or settling of rocks (displacement) somewhere in the rather weak, unstable sections of the earth's crust. Such shock waves occur when stresses build up to the point where the rocks break. Most major earthquakes probably result from faulting although minor shocks may come from volcanic activity, landslides, and subsidence. People have also been responsible for some earthquakes. For example, the injection of

Fig. 7-24. Landforms associated with vulcanism.

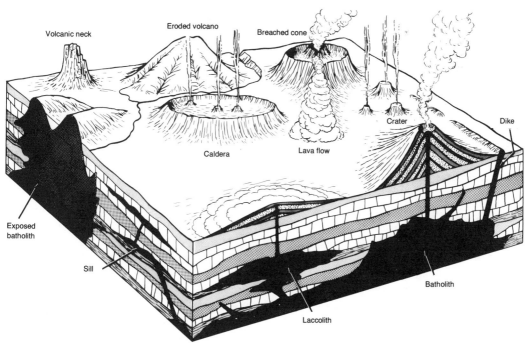

Fig. 7-25. Wreckage on 4th Street, Anchorage, Alaska, after the severe earthquake of March 1964.

water or liquid wastes into specially con-structed deep wells or old mines has lubri-cated faults, reducing their resistance to stress and triggering small earthquakes. The use of Amchitka Island in the Aleutians as an underground atomic testing area has been a source of concern because the island lies close to the Aleutian thrust fault. Many peo-ple view testing in such areas as very danger-ous because of our lack of knowledge about faulting and its relation to earthquakes.

The Richter scale, developed by Charles Richter in 1935, measures the amount of energy released at the earthquake's origin on a scale from 1 to 9. The scale is logarithmic, that is, a quake with a magnitude of 6 is approximately 31.6 times greater than one of 5 and 1000 times greater than one of 4. A major quake has a magnitude of 7 or more. The largest recorded quake probably took place near Anchorage, Alaska on 27 March 1964; it measured 8.4 on the Richter scale.

In some instances the displacement caused by earthquakes can be seen on the surface of the crust, as illustrated in Fig. 7-25 (the Alaskan quake of 1964). Usually, however, displacement occurs below ground. Most of the hundreds of thousands of earth-quakes that occur each year pass unnoticed because they are slight or because they occur in uninhabited areas or beneath the oceans.

Though earthquakes are commonly

thought of in connection with their immediate destructive power, they are also responsible for other kinds of destruction. For example, the tremendous and almost instantaneous destruction caused by earthquakes often gives rise to widespread fires. In fact in the great Tokyo earthquake of 1923, when an estimated 100 000 people died, most of them met their death by fire. In addition, earthquakes under or adjacent to the oceans may set in motion enormous waves (tsunamis) that travel at speeds of up to 600 km/h. Striking a coastline several metres above the high tide mark, they can cause severe damage to low-lying areas thousands of kilometres from the source of the quake. The type of destruction caused by an earthquake is also related to the characteristics of the earth's surface where the quake occurs. For example, in the 1964 Alaskan earthquake the majority of the damage occurred on the weaker clays, particularly in sloping areas. These surfaces subsided and were pulled apart, causing roads and buildings to be tilted and broken. Buildings built on solid rock suffered little damage.

STUDY 7-6

1 Make a cross section diagram of the Black Hills of South Dakota (Fig. 7-23) and explain how this landscape was created.

2 Referring to Fig. 7-24, write a definition for each of the examples of intrusive vulcanism shown.

3 Referring to Fig. 7-9, which shows the distribution of earthquakes, identify the earth's principal earthquake zones and comment on their relationship with the movement of the earth's plates.

4 Both volcanoes and earthquakes constitute major environmental hazards, with earthquakes being the more serious. Many people are studying these natural phenomena with a view to reducing their catastrophic effects. In simple terms this work has three main aspects: (1) prediction of the time of eruption or quake, (2) preparation for the hazard by precautions such as erecting buildings strong enough to withstand an earthquake or removing people from areas of potential volcanic eruption, and (3) prevention of volcanoes and earthquakes. These are important subjects described in a number of books listed in the bibliography at the end of this book. Do you feel that any one of the three aspects above warrants priority? Why are most people reluctant to support or be advised by such research?

[1]Glenn T. Trewartha, et al., Fundamentals of Physical Geography (2nd ed.; New York: McGraw Hill Book Company, 1961), p. 42. Copyright 1961 the McGraw-Hill Book Company, Inc. Used with the permission of McGraw-Hill Book Company.

[2]Based on R.S. Dietz, 'Geosynclines, Mountains and Continent-Building', Scientific America, Vol. 226, No. 3 (March 1972).

[3]Vernor C., et al., Elements of Geography (4th ed.; New York: McGraw-Hill Book Company, 1957), p. 217. Copyright 1957 the McGraw-Hill Book Company, Inc. Used with permission of McGraw-Hill Book Company.

8/ WEATHERING AND THE FORCES OF GRADATION

We have completed a review of the earth's crust and in particular the tectonic forces that derive their energy from the earth's interior and move, bend, break or otherwise alter its surface. The object of this chapter is to examine another set of forces known as the gradational forces. Driven by solar energy, they are responsible for producing the distinctive features that make up the earth's landforms.

The irregularities on the earth's surface created by folding, faulting, and vulcanism are constantly being modified by gradation. The agents of gradation include running water, gravity, glacial ice, wind, and waves. Gradation occurs when surface rock disintegrates or decomposes and is transported by gradational agents to other areas, where it is then deposited. Through gradation the higher parts of the earth's surface are lowered and the lower sections are raised. Ultimately gradation reduces or eliminates differences in relief, tending to produce low-lying, level land. If all tectonic activity were to cease, the gradational forces would, over an immense period of time, reduce the land to an almost level plain. Of course this never happens. Because tectonic activity is a continuous though erratic process, few areas of the earth

are ever reduced to a level surface. However, all parts of the earth's land surface are affected by the gradational agents. In fact their influence is so great that at least the superficial characteristics of all landforms are a result of gradational rather than tectonic activity.

WEATHERING

Weathering is the first step in gradation and includes all processes that decompose and disintegrate rock. It produces a finely fragmented layer of rock particles (known as *regolith*) that the agents of gradation can then carry away. Although weathering processes are extremely complex and in some cases not clearly understood, they can be subdivided into two groups—mechanical and chemical. Both types of weathering occur almost everywhere on the earth's land surface. They prepare materials for transportation and deposition by the gradational agents and act as the principal agents in soil formation. Seldom, however, are they responsible for large and distinctive landforms.

Mechanical Weathering

This term is used to describe the breaking up of rock into smaller fragments without any apparent chemical change in the rock itself. Mechanical weathering occurs most frequently in bedrock that has been highly fractured. Cracks and crevices are exposed to the atmosphere and water seeps into them. When the water freezes, it exerts a powerful force that breaks up the rock. In arid areas salt crystals also exert pressure on the rocks. The evaporation of water in the cracks of porous rocks such as sandstone leaves behind tiny salt crystals. As these crystals grow the rock breaks apart, leaving the rock fragments exposed to wind erosion. The destructive effect of salt crystallization is also seen on brick or concrete buildings that are in contact with moist soil or with the salt used in cities to combat winter ice.

The importance of heating and cooling of rocks as a cause of weathering is in dispute. Nevertheless many earth scientists believe that repeated contraction and expansion causes the outer layers of rock to fracture or disintegrate. The breaking away of rock sheets or scales is thought to be at least partly a result of expansion and contraction. This process is called *exfoliation*.

Finally, the roots of plants may break up rock as they grow and spread in search of water and nourishment. Anyone who has seen the damage that the roots of trees (or even weeds) can do to a road or a cement sidewalk in a relatively short time will appreciate the effect they can have over many centuries.

Mechanical weathering also results from the abrasive action of previously weathered fragments carried along by water, wind, or ice.

Chemical Weathering

Chemical weathering occurs when minerals in the rocks of the earth's crust come in contact and react with carbon dioxide, water, or oxygen. The chemical reactions usually cause the minerals in the rock to swell, thus exerting pressure that may weaken the chemical bonds that hold the rock together, making it more susceptible to decomposition. In addition, such changes in the composition of certain minerals often produce new minerals that are softer or more soluble than the original ones. This also weakens the rock, causing decomposition. For example, potassium feldspar—a common mineral found in granite—changes to clay in reaction with water and carbon dioxide. Clay is a soft mineral that tends to absorb water, thus swelling and causing the rock to disintegrate. This type of chemical weathering is referred to as *carbonation*.

Another type of chemical weathering known as *oxidation*, results from the reaction caused by contact between iron-bearing minerals and the oxygen dissolved in water. This familiar phenomenon, called rusting, causes the iron to change to iron oxide, which is a softer material than iron. Because iron is a very abundant element in the rocks of the crust, rusting is a common cause of weathering.

Solution is another type of chemical weathering. It occurs when dissolved carbon dioxide from the atmosphere and organic acids from decaying plant matter are added to rainwater, forming a weak carbonic acid. The acid reacts with such basic rocks as limestone to dissolve certain minerals (calcite, for example) and to carry them away in solution, as the following equation illustrates:

$$CaCO_3 \quad + \quad H_2CO_3 \quad = \quad Ca(HCO_3)_2$$

| Calcite | carbonic acid | calcium bicarbonate |

Because calcium bicarbonate is soluble it may then be carried away in solution; since calcite is the chief constituent of limestone, the rock will gradually dissolve. Solution is an important process in landform development in the many areas where limestone occurs, and is further discussed on page 218.

Factors Influencing Weathering

Weathering operates very slowly, even under the most favourable conditions. Over millions of years, however, it creates a layer of regolith of varying depth over the unweathered bedrock. The depth of the regolith in any location depends on several factors including climate, type of rock, and slope.

CLIMATE

The amount and type of weathering vary according to climatic conditions, as indicated in Fig. 8-1. Weathering is particularly noticeable, however, in areas with high temperatures and abundant precipitation. In regions that are very dry or very cold the layer of regolith is mainly the result of mechanical processes and is normally shallow. Even within a relatively small area where differences in the amount and type of weathering result from local variations in climate, these variations are largely a result of differences in factors such as exposure to the sun and moisture-bearing winds. For example, slopes facing south in the northern hemisphere are less steep than slopes facing north—probably because the northern slopes have longer periods of snow cover, fewer days of freeze and thaw, and retain soil moisture better. These factors contribute to a slower rate of weathering.

TYPE OF ROCK

The rate of weathering also depends on the type of rock. Sedimentary rocks, as their process of formation suggests, are the least resistant to weathering. They are particularly liable to mechanical weathering. (However,

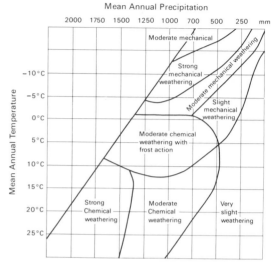

Fig. 8-1. The relative importance of different types of weathering under different temperature and precipitation conditions.

sedimentary rocks containing silica—for example, some of the conglomerates or sandstones—are among the most resistant of all rocks.) The reaction to weathering depends largely on the strength of the cementing material that binds the particles together. Some sedimentary rocks also react differently to climate. Limestone is vulnerable to chemical weathering and thus tends to decompose fairly rapidly in humid regions. In arid areas, however, it is among the most durable of the sedimentary rocks and is often found capping high ridges. Sandstone is usually quite resistant to weathering in both humid and arid areas; shale—another common sedimentary rock—weathers rapidly in both arid and humid areas. It should be noted that such comparisons are relative because weathering of all kinds proceeds far more slowly in dry areas.

Though igneous rocks are generally resistant to weathering, they are more susceptible to chemical than to mechanical weathering.

A HUMID

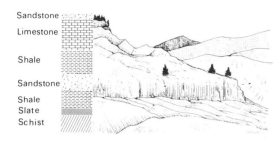

Sandstone
Limestone
Shale
Sandstone
Shale
Slate
Schist

B ARID

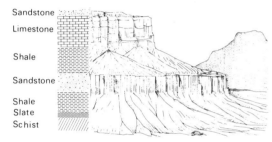

Sandstone
Limestone
Shale
Sandstone
Shale
Slate
Schist

Fig. 8-2. The lower diagram illustrates the effect of differential weathering in an arid climate. The upper diagram shows how weathering would affect the same combination of rock types in a humid climate.

Slow disintegration occurs as minerals are chemically altered or as water is absorbed, causing the rock to swell and eventually to crumble.

Metamorphic rocks, particularly those composed of quartzite, are the most resistant of all rocks. Since they are found on the surface less often than igneous and sedimentary rocks, their weathering characteristics are not as important.

Differential weathering becomes obvious where different types of rocks are interbedded, as shown in Fig. 8-2. The presence of a resistant rock (*cap rock*) above beds of more

easily weathered rocks will tend to protect the less resistant rocks, forming an *escarpment* (also known as a *scarp slope*).

SLOPE

The depth of the regolith depends on the degree of slope as well as on the rate of weathering. On steep slopes gravity causes fragmented rock to be carried away as quickly as it is formed, leaving behind only large, coarse stones or even exposed unweathered rock. On gentle slopes, layers of weathered material become progressively deeper on the lower part of the slopes. On level surfaces considerable depths of regolith are usually found; in tropical areas, depths of 50 m or more of weathered material above bedrock are not uncommon.

STUDY 8-1

1 Explain the relationship between weathering and climate shown in Fig. 8-1.
2 Because different kinds of rock in the same area weather at different rates, surface irregularities are produced. Discuss the effects of differential weathering in the landscapes portrayed in Fig. 8-2. Fig. 8-11 provides another interesting illustration of this phenomenon.

MASS WASTING

The importance of gravity in creating landforms is often overlooked. It exerts a continual downward pull on the weathered particles (regolith) that cover much of the land. If anything disturbs the balance of these particles, they will move. Such movements are referred to as *mass movement* or *mass wasting* even though in many instances only single particles are involved. Because gravity continually affects all sloping surfaces, the total amount of material moved over a long period of time will alter the land significantly.

Whenever the regolith on a sloping surface

is disturbed, gravity may cause the regolith to move downslope—either single particles or huge masses of rock may move anywhere from a few millimetres to thousands of metres. Such mass wasting is one of the principal ways in which weathered material is carried downslope. Consequently gravity may be regarded as one of the forces of gradation.

Various factors can disturb the balance of the materials on a slope, thereby contributing to mass wasting. Some of the more important are: wetting and drying; freezing and thawing; thermal expansion and contraction; earthquakes; and, on a smaller scale, the activities of people and animals. The actual amount of material that will be shifted by gravity—and the speed at which it will move —can be related to several additional factors. These include the climate of the area, the types of underlying bedrock, the type of vegetation, the depth and other characteristics of the regolith itself, and of course the steepness of the slope.

STUDY 8-2

1 Examine the factors listed above and note how each may contribute to the downslope movement of material by gravity.
2 Explain why the work of gravity is often underestimated, indicating how mass wasting may be as important an agent of gradation as running water.

Landslides

Mass movements involving the *rapid* movement of rock particles are called landslides. The simplest landslide, which occurs in most mountainous regions, involves the downhill movement of rock particles, singly or in small numbers. The pile of debris that results from it at the base of almost any steep slope is known as *talus* or *scree* (Fig. 8-3).

A more spectacular type of landslide occurs when a substantial mass of rock or soil breaks away and moves rapidly downhill. The causes of these slides are numerous. Among the more important are the existence of weak or stratified bedrock (particularly where the fractures in the rock are inclined) and the presence of an unusually thick layer of regolith. A slope with either of these characteristics may be unstable. A large part of it could be set in downhill motion if the slope's balance is disturbed by a river undercutting it, saturation with water, earthquakes, or even disturbances related to mining or highway construction.

On 31 May 1970 an earthquake measuring 7.7 on the Richter scale (see p. 190) occurred in the coastal mountains of Peru. It triggered a devastating rock and ice avalanche on the north summit of Mt. Huascarán (6665 m), where an 800 m strip of the mountain broke off near the top and fell into the valley below. A geophysicist who saw the landslide recorded that the earthquake began at 23 minutes 28 seconds past 3 p.m., and that by 25 minutes 25 seconds past 3 p.m. the avalanche had covered 16 km and had almost completely destroyed the agricultural towns of Yungay and Ranrachirca as well as ten small villages in the valley below the mountain. An estimated 25 000 people were killed. Though this was the worst disaster of its kind in the Andean region of South America, there have been many other destructive landslides triggered by earthquakes that can be linked with the subduction zone (Peru-Chile Trench) off the coast of South America (see Fig. 7-9). Much of the coastal area of Peru is similarly vulnerable. However any attempt to depopulate these areas would be unrealistic because of their agricultural productivity. Security measures are essential if people are going to live with such environmental hazards. These measures might include the construction of earthquake-resistant buildings, the re-siting of settlements that are now in particularly vulnerable locations, and the rebuilding of towns with

Fig. 8-3. Accumulations of ice-shattered rocks called talus or scree can be seen at the base of these steep slopes in the Wallowa Mountains of eastern Oregon.

wide streets. Whether a developing nation such as Peru can undertake such measures is another question.

Most landslides are variants of slumps. Both rock and regolith break away from a slope along a concave surface. As the slump moves, the upper part of the landslide drops below the normal ground level while the lower part of the landslide is forced above the former level of the land's surface. Figures 8-4 and 8-5 show slump-type landslides. It was a slump landslide that kept the Panama Canal

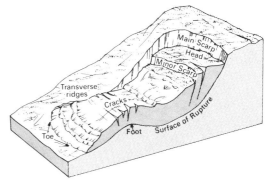

Fig. 8-4. The principal parts of a slump-type landslide.

Fig. 8-5. A slump-type landslide, north of Tepeköy, Turkey, triggered by an earthquake in August 1966.

closed for most of the period between 1914 and 1920. In a part of the Canal known as the Culebra Cut, slides of 180 000 to 650 000 m³ occurred almost weekly, blocking the canal and—in some instances— heaving up the bottom of the canal as much as 10 m.

Earthflows

Earthflows are a slow mass movement that usually occur on steeply sloping, grass-covered hills. Figure 8-6 shows various stages in this type of movement. Earthflows are caused by the saturation of the ground during heavy rainstorms or periods of spring thaw. Water acts as a lubricant, increasing the weight of the regolith and reducing the cohesiveness of the slope. Earthflows can be easily recognized from scars in the earth or from displaced vegetation.

Rapid and localized earth flows are referred to as *mudflows*. Because of their great velocity and density, they can transport large objects for short distances. In mountainous and subhumid areas, where they are common, mudflows have been known to bury or dislodge buildings, automobiles, or large trees that lie in their path.

The earthflows described above usually affect particular areas of a slope. A slightly different type of earthflow, known as *solifluction*, occurs in areas where the entire regolith is saturated with water and the whole earth mass—or a large part of it—moves slowly downhill. This is particularly common in areas of high latitude or high altitude, where only the upper part of the surface material thaws during the summer and permanently frozen ground (*permafrost*) lies a short distance below the surface. When a

thin, water-saturated layer rests on top of a frozen base that is not completely level, the upper layer tends to ooze downhill. Solifluction thus creates small, irregular wrinkles on the land surface.

Creep

Creep is a slow mass movement of the surface of the regolith. It involves the downhill movement of all loose particles, but produces no obvious landforms. Normally creep occurs too slowly to be perceived. However, the cumulative effect of the slow movement of surface material downhill on a slope alters the land substantially.

Creep is probably the most important

Fig. 8-6. Active earthflows on the west side of Pleitito Canyon in the San Emigdio Mountains of California. The difference in altitude in the area photographed is about 360 m.

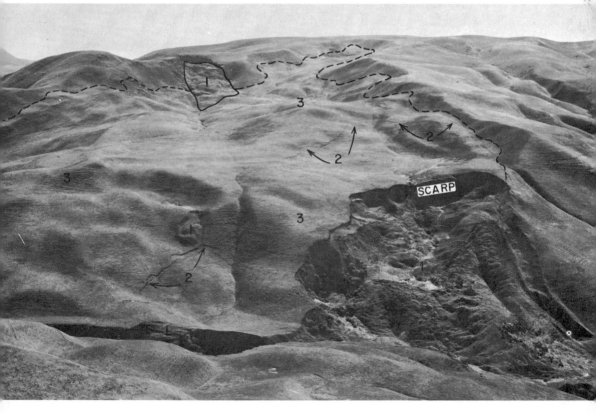

example of the way gravity reduces irregularities on the face of the land. It occurs in several ways. For example, the formation of frost crystals lifts up soil particles and, when frost melts, the particles will probably be shifted slightly downhill. Creep can also be set in motion by the expansion and contraction of rock fragments owing to changes in temperature. The growth of vegetation may also cause particles to move downward. Eventually such particles will either accumulate in depressions or be carried away by other gradational agents.

<hr>

STUDY 8-3

1 From Fig. 8-6 identify three or four examples of earthflows in process. There are several small scarps in the same photograph, each of which indicates the beginning of a new earthflow. Identify these and describe how the earthflows will develop. Could mass movement of this sort be prevented, and under what circumstances might this be necessary.

2 Figure 8-6 shows where a much larger slump occurred some time ago. Identify the scarp left when the surface material broke away and also determine the extent of this earthflow.

3 Explain why creep is often described as the most important form of mass movement. Why is the impact of creep greater in landscapes where people have altered or destroyed the natural vegetation?

4 Discuss the security measures noted in connection with the Peruvian landslide and suggest some alternatives. See question 4, Study 7-6, p. 191.

<hr>

GRADATION BY RUNNING WATER

Running water (technically known as runoff) and ground water are two processes of the hydrologic cycle described in Chapter 2 (pp. 21 to 22). This cycle and the additional material on 'Waters of the Land' should be reviewed before proceeding with this section.

In most parts of the world, running water is the major gradational agent. Almost every area receives precipitation, some of which becomes runoff. This runoff always seeks a lower level in streams and rivers, in the process sculpturing the landscape into valleys and intervening hills. Over a long period of time runoff slowly levels the entire surface. Given an endless stretch of time and no tectonic activity, running water probably could wear down the highest mountain range of the toughest rock to an almost featureless plain.

Though gradation by runoff is our main concern here, it is important to remember that both surface and groundwater runoff are basic natural resources. They provide most of the water used for domestic and industrial purposes and also water for irrigation, hydroelectric power, inland transportation, and recreation. Runoff has been a misused resource. Water pollution is one example. Thus a knowledge of runoff is important to anyone who studies the use and misuse of water resources.

Runoff

All runoff begins with precipitation, and Fig. 8-7 illustrates the various factors affecting runoff during a period of precipitation. Most precipitation first moves over the surface of the earth in a continuous film or a series of small threads of water flowing between plants or small surface irregularities. This is called *overland flow*. At the bottom of slopes this flow becomes concentrated in *stream flow*, which grows progressively larger towards the mouth of the river system.

Overland flow inevitably results in erosion. Under normal conditions the rate of erosion is slow and the loss of soil particles from the surface is balanced by the formation of new soil materials below the surface. The rate of erosion may be accelerated on sloping terrain where people have destroyed or altered the natural vegetation. This process begins when overland flow causes the movement of rock or soil particles across the entire surface. (This is known as sheet erosion.) Soon rills form in the cracks and crevices of the land, gradually deepening these cracks to form miniature valleys. (This is called rill erosion.) If the runoff is great enough and the slope sufficiently steep the rills will develop into steep V-shaped gullies (referred to as gullying).

STUDY 8-4

1 Referring to Fig. 8-7, explain the sequence of events from the beginning of a rainstorm on a sloping surface to the discharge of overland flow into a river or stream.

2 Explain how the rate of overland flow will vary according to the following factors: the amount, intensity, and duration of precipitation; the amount and kind of vegetation; the type of soil; the type of structure of bedrock; and the steepness of the slope.

3 Explain how accelerated erosion may be caused and suggest measures that might stop this destructive process once it has begun. Can the erosion caused by stream flow also be accelerated as a result of human interference in the natural landscape? Explain.

The Work of Rivers

Runoff is a major factor in land sculpture whether as overland or stream flow. Through erosion, transportation, and deposition it has affected virtually every part of the earth's land surface.

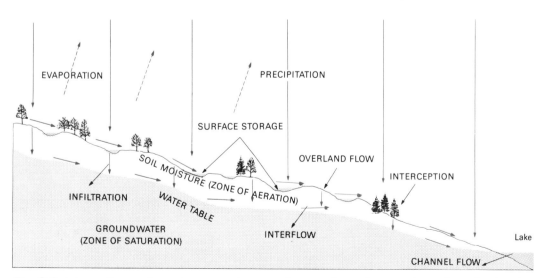

Fig. 8-7. Interception, surface storage, and evaporation vary from place to place, thus affecting the amount of precipitation that eventually becomes ground water or overland flow.

EROSION

During periods of heavy rainfall overland flow may carry particles of soil dislodged from exposed positions of the earth's surface by the impact of raindrops or by the velocity of the moving water. This runoff picks up more particles by the scouring action of the particles already being carried along. The water comes together in small depressions, forms rills, and finally joins permanent streams (stream flow). Both the force of the moving water on the sides of the stream's banks and the scouring action of the particles already being carried along cause more material to be picked up. Stream flow thus causes the formation of river systems. As erosion continues, all the valleys of a river system are deepened and widened, and the entire landscape is lowered.

The three principal means of erosion are: *hydraulic action*, the removal of loose particles by the force of running water; *corrasion* (or abrasion), the mechanical wearing away of the land by the grinding action of particles carried by water; and *corrosion*, the gradual disintegration of rocks through chemical reactions between water and certain minerals in the rock.

TRANSPORTATION

Stones and large boulders may be rolled or bounced along a river bed, whereas smaller particles are carried along in suspension and in solution. The ability of a river to transport eroded materials depends largely on its speed and volume. As the speed increases, the erosive power becomes greater and more material can be transported. Ability to transport increases greatly with only a slight increase in speed. For example, the doubling of a river's speed will increase its transporting ability sixty-four times. This explains the enormous destructive power of rivers in flood, when both the speed and the volume of water have increased substantially.

It has been estimated that each year rivers in the United States carry 245×10^6 t of dissolved material and 462×10^6 t of solid material from the surface of the continent to the ocean. If all of this material were removed uniformly it would mean a lowering of the entire surface of the United States by about 30 cm every 10 000 years.

DEPOSITION

Deposition occurs when the carrying capacity of a river is reduced. A decrease in gradient, for example where a swift mountain stream flows on to a plain, may cause deposition. It also takes place when the speed of a river is reduced on entering a large body of water. Deposition may result from an increase in the load of material to transport. This occurs, for example, when very muddy water from a fast-flowing tributary joins a larger and more slowly moving river. Rivers that have their source in humid areas may have their carrying capacity reduced as they pass through sub-humid areas because of water loss owing to evaporation. Though deposition normally increases downstream, it may occur at almost any point in a river's course. Deposition often occurs on the inside bend of a river because velocity is less there than on an outside bend (Fig. 8-9).

Material deposited by rivers and streams is known as *alluvium*. The fragments vary in mass and size. Coarse gravel and sands are usually deposited first, and finer silts and clays are carried farther before being laid down. During periods of normal flow deposition occurs only on the bed of a river, but during periods of flooding it will extend as far as the flood waters—perhaps across the entire floor of a valley. Most slowly flowing rivers change their course many times. This process not only widens the valley but causes the floor of the valley to be filled with an increasingly wide and deep deposit of alluvium. These almost flat surfaces are known as

Fig. 8-8. Severe rill erosion.

floodplains. The Ganges River in India and the Hwang Ho (Yellow River) in China have very large floodplains.

Stages of Gradation

Ultimately rivers wear down the land in their drainage basins. The amount of lowering depends on the *base level* of the drainage system concerned, that is, the altitude at any given point in the drainage system below which the river would cease to flow. The true base level of all drainage systems is determined by the level of the ocean. However some rivers, such as those flowing into the Great Lakes (which are above sea level), or those flowing into the Caspian Sea (which is below sea level), have a temporary base level other than sea level. While the lowest point of a river's base level is the level of the lake or sea into which the main river flows, the base level rises gently towards the upper reaches of a drainage basin. A drainage system will lower the land it drains to base level but never below, because below base level the remaining slope would not be sufficiently inclined to enable rivers to flow and carry away runoff.

As the land in a drainage basin is lowered, certain stages of gradation can be seen. These stages are referred to as early (youthful), middle (mature), and late (old), depending on the extent to which the land has been

Fig. 8-9. Map and cross-section to illustrate deposition and the formation of meanders on the lower Mississippi River.

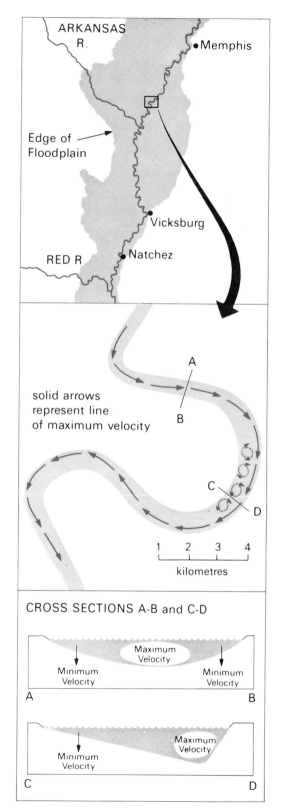

solid arrows represent line of maximum velocity

CROSS SECTIONS A-B and C-D

lowered in relation to base level. Since the volume of water carried towards the mouth of a river is normally much greater than in the upper reaches of the drainage system, the gradational power will be greatest near the mouth. Thus lowering will normally have progressed much more near the mouth. It is common for a river to have reached base level and to be in a late stage near its mouth. Farther upstream, downcutting may still be proceeding, and the main river, together with its tributaries, will be in a middle stage. Closer to the sources, or at the edge of the drainage basin, the many small tributaries may not have substantially lowered the land, and they will be in an early stage.

This classification of rivers according to stages of gradation is a common and useful form of generalization and is referred to in subsequent sections of this chapter. However, it is important to emphasize that the idealized sequence of stages presented here is seldom as clear-cut in reality. This would be apparent to anyone making an actual study of a number of river systems.

Valley Development

The development of a valley is a result of three processes—valley deepening, valley widening, and valley lengthening. As previously noted, valley deepening is a result of river erosion. Valley widening, however, occurs in a number of ways. Overland flow causes materials to be carried down from the sides of the valley and leads to the development of gullies that become tributaries draining into the main stream. Lateral erosion by the moving water in the stream itself causes the stream banks to be undercut so that eventually they slump or slide into the river. The downslope movement of particles, which also contributes to the widening of a valley, can be caused by water and mass wasting.

The most important process in valley lengthening is headward erosion, that is, the development of valleys in an upslope direction. This process can often be seen on a hillside where the vegetation cover has been removed. On the unprotected surface small gullies will begin to form at the base of the slope and, unless checked by some protective measure, will gradually deepen and extend up the slope (Fig. 8-8). In the same way, but more slowly, river systems will deepen existing tributaries and extend new ones towards the source regions of the river system. Valleys may also be lengthened at the river mouth by a lowering of the sea level, the uplift of the land through warping, or the extension of deltas. For example, many of the rivers along the southeastern seaboard of the United States now flow across a land surface that was once part of the continental shelf.

At the transition between youth and maturity in the development of a river system, the process of valley deepening causes streams to slope just steeply enough to transport eroded materials. Streams at this stage are called *at grade* or *graded*. The longitudinal profile of such a stream is known as a *profile of equilibrium*. Such a profile for any major river is seldom a regular one but rather consists of a number of segments, each slightly different from the other. The segments are usually steeper towards the source, but not in every instance. For example, the profile of a river downstream from a confluence is often steeper than the profile of the main river above the confluence. This occurs because a tributary brings a heavy load of eroded materials into the main river, thereby necessitating a steeper gradient in the main river in order that the materials can be transported. The achievement of a graded profile does not mean that the river system has attained its lowest gradient, but that further valley deepening will occur only very slowly. The profile of the graded stream will represent the local base level for the tributaries that drain into the graded stream.

More information about valley development can be obtained by examining cross sections of valleys. In the youthful stages V-shaped profiles occur most frequently, suggesting that vertical erosion or valley deepening is the most important gradational process. But valley widening is also occurring and in some cases may be of almost equal importance. After a river has achieved a profile of equilibrium, lateral erosion becomes the dominant process. It is reflected in the meandering of the river (meanders are discussed on p. 213 of this chapter) and in the development of a floodplain. In its early stages the floodplain is very narrow and the valley floor (*erosional surface*) is covered by only a thin veneer of alluvium. As the valley continues to develop into old age it becomes very wide—certainly several times wider than the actual belt occupied by the meandering river—and the depth of alluvium becomes much thicker. For some of the largest rivers, such as the Mississippi, the alluvium is over a hundred metres deep in places.

Though it is convenient to describe valley development by the stages of gradation, it is important to remember that the stages do not always occur. Some reasons for the exceptions are described below.

Some valleys are cut so that they are narrow and V-shaped in cross section, while others are steep-walled and somewhat U-shaped. The traditional view has been that such shapes are characteristic of the stage of the erosion cycle of the stream. Hence, a V-shaped valley was taken to mean a stream in youth, while a valley which was broad, wide-bottomed, and steep-walled was mature or old. However, rather than indicating a stage of erosion, valley shape is often the result of the interaction of climate, available relief, rock type, and geologic structure. Climate directly influences both the volume of flow in a stream and vegetative cover on the surface. Vegetation partly controls the amount of runoff, as well as supplying a protective cover to the surface. Its effect is such that in extremely rainy, tropical areas stream valleys are deep and narrow because the dense vegetation prohibits wall erosion. In temperate, moist regions wide walls are smoothed to a gentler slope by creep and soil wash, giving a typical V-shaped valley. On the other hand, in arid regions where the ground surface is bare and exposed, stream valleys are often steep-walled and flat-bottomed, resulting in a U-shape.[1]

Rejuvenation

A river that has reached a profile of equilibrium will begin to erode downwards after a change in base level or a significant increase in the amount of water or the load carried by the river. Such erosion will re-establish youthful features. This process is called *rejuvenation*. There are a number of ways in which it may occur.

Eustatic rejuvenation results from a lowering of the sea level. During the ice ages, water was evaporated from the oceans, lowering the base level of all the world's river systems and rejuvenating at least the lower courses of the rivers. Eustatic rejuvenation is also thought to be slowly occurring as a result of sea-floor spreading (p. 171). As the continents have been spreading apart, the floors of the ocean basins have been sinking. This causes an extremely slow but steady fall in world sea levels.

Dynamic rejuvenation occurs when tectonic activity causes stream gradients to be steepened and the stream starts downcutting again. Eventually this process re-establishes a graded profile.

Static rejuvenation occurs if the load car-

ried by the river is substantially reduced or if the average amount of rainfall over the drainage basin is substantially increased. In either case a graded stream will start downcutting again.

Several landform features are associated with rejuvenation. As the river begins downcutting, the old valley floor remains as a flat area separated from the new level by a scarp. This is called a *river terrace*. The terrace lies some distance above the new level of the downcutting river, and the new river no longer flows on it, even during periods of flood. *Entrenched meanders* also result from rejuvenation. They are formed when rejuvenation affects a river that has been meandering on a floodplain. The river starts downcutting again to re-establish grade, while maintaining its winding course. The old meanders remain in existence, but are found in V-shaped, youthful valleys.

STUDY 8-5

1 Figure 8-10 shows three stages in the development of a landscape by running water. Each sketch and its corresponding contour map represent a small portion of a drainage system.

a Draw a simple cross section of each stage to show the width, depth, and shape of the valleys.

b Write brief notes on the degree of tributary development in each stage of development, referring to the ability of the river system to drain the land. Why is the river system unable to drain the land completely at certain stages?

Fig. 8-10. These three block diagrams and their corresponding contour maps show some of the characteristics that might be found in the different stages of river development.

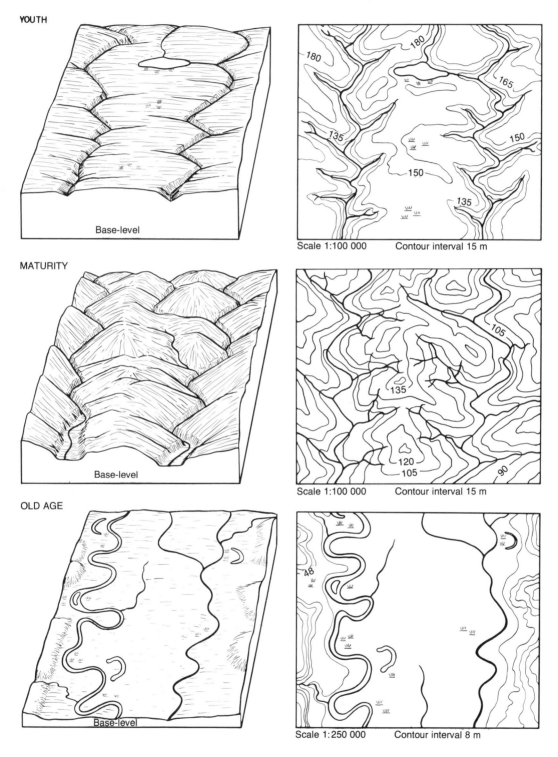

YOUTH

Base-level

Scale 1:100 000   Contour interval 15 m

MATURITY

Base-level

Scale 1:100 000   Contour interval 15 m

OLD AGE

Base-level

Scale 1:250 000   Contour interval 8 m

c For each stage calculate the average slope of the river bed in metres per kilometre and then comment on the differences.

d Describe the degree to which the interfluves have been dissected and lowered in each stage.

Using the answers to the questions, organize a description, including a sketch, of each of the stages represented in Fig. 8-10. The description should contain references to the processes of erosion, transportation, and deposition.

2 How do we know when valleys are a product of erosion by water? Can valleys be formed in any other way?

3 Using maps of several different scales (a continental map in an atlas and a topographic map) draw profiles of the courses of several rivers. Relate the profiles you draw to the idealized states of erosion.

4 Assume the old-age stream in Fig. 8-10 has been rejuvenated. Make a contour sketch to illustrate the terraces and entrenched meanders that might have developed.

Landscapes Dissected by Rivers

Youthful landscapes may be of two types. First, many are youthful because there has not been enough time for tributaries to dissect the surface. Such surfaces are characterized by interfluves that have not been greatly dissected, widely spaced river valleys, and a limited number of tributaries. Second, some areas would seem to remain almost permanently youthful. The latter often result from a combination of several factors: a flat surface, a complete vegetation cover, low precipitation, and a highly permeable subsurface. Much of the rainwater that falls on such areas sinks into the ground and eventually flows as ground water into the main channels. As a consequence tributary development is minimal, the interfluves remain relatively undissected, and only the main channels slowly deepen and widen. Eventually

such a landscape may pass from youth to old age without going through the intermediate stage. Large areas of the 'high plains' of the western United States and Canada exhibit such long-term youthfulness.

Mature landscapes differ from their youthful counterparts in that tributaries have thoroughly dissected their interfluves. The resulting landscape is generally one in which level land is at a minimum; in fact such a landscape is made up almost entirely of valley slopes. The various characteristics of the underlying materials—including their resistance to erosion, permeability, the amount and type of precipitation and vegetation, and the nature of the initial surface—all affect the features of a mature landscape. An example of the effect of underlying rock structure on terrain features is shown in Fig. 8-11. Here formerly horizontal layers of different types of sedimentary rock have been warped. After the drainage system developed it eroded most quickly over the least resistant rock types. Gradually it developed a distinctive pattern of strips of generally flat land separated by low ridges that slope much more steeply on one side than on the other. Such ridges are known as *cuestas*. Examples of irregular plains with cuestas include the eastern part of the Paris Basin (Fig. 7-16), the Weald of southeastern England, the Niagara escarpment of southern Ontario, and a number of areas along the eastern and southern seaboards of the United States. Though cuestas are only minor features (seldom more than 100 m high) they often indicate different types of rock and hence different characteristics of soil and drainage. This information may be of great assistance in understanding variations in land use from one side of a cuesta to the other.

It is very hazardous to suggest a cycle of erosion and a sequence of stages in both youthful and maturely dissected landscapes because so many different situations exist. Once the processes affecting surface-water

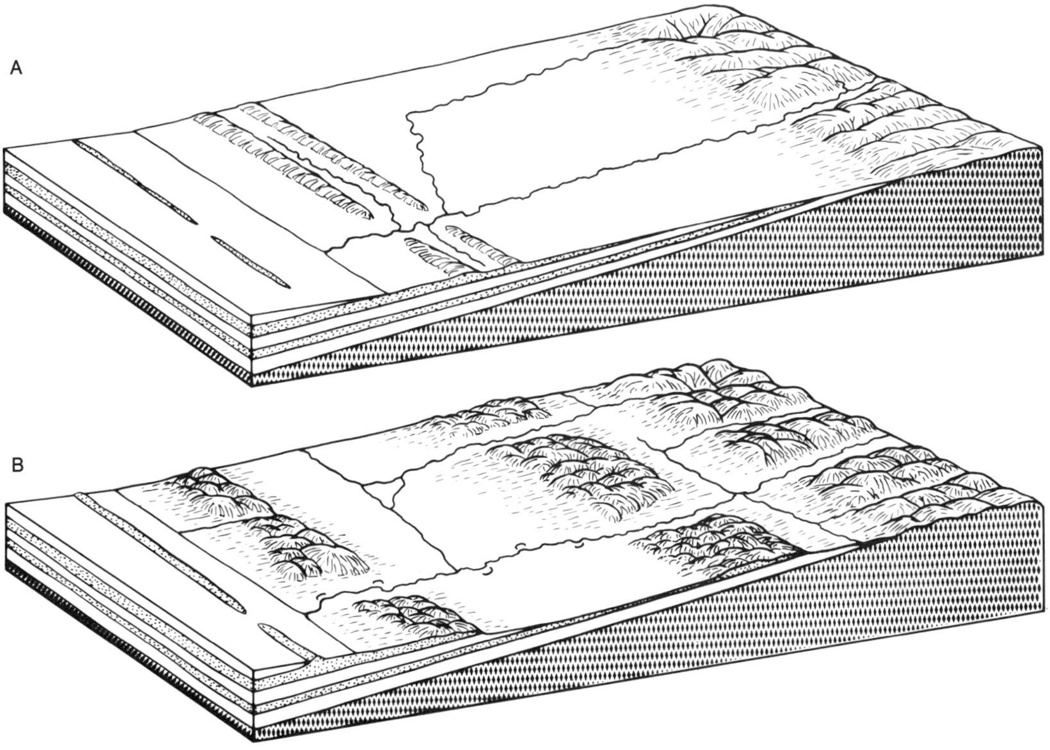

A

B

Fig. 8-11. A and B represent the same area separated by a substantial period of time and illustrate the development of cuestas on a coastal plain. The shaded sedimentary strata (e.g. sandstone) are more resistant to erosion than unshaded strata (e.g. shale). Cuestas usually consist of both a dip and a scarp slope. Identify these in B and explain how they were formed.

gradation are understood, it is essential to apply this knowledge to an analysis of specific landscapes. Study 8-6 is the basis for such an analysis.

STUDY 8-6

1 Examine the characteristics of the landscape in Figs. 8-12 and 8-13 (using a stereoscope) and make cross-section diagrams of it. Discuss the stage of development of the landscape and suggest reasons for the flat and largely undissected interfluves.

2 Using cross-section diagrams analyse the landform characteristics of the area on Fig. 8-14. What stage of development is it in? Compare this landscape with that shown on Figs. 8-12 and 8-13 and comment on their similarities and differences. (Your answer will require an understanding of the climatic differences between those two areas.)

3 Explain how the cuesta landscape illustrated in Fig. 8-11 was formed. What factors would determine the height of these cuestas?

4 Examine Fig. 8-15 carefully. Using cross-section diagrams, explain the stage of devel-

Fig. 8-12. Codroy, Newfoundland (part of), Canada National Topographic Series. Scale 1:50 000. Contour Interval 15 m. Fig. 8-13. This stereo photograph shows part of the area on Fig. 8-12.

opment of this landscape. In your explanation note the characteristics of the Sabine River, its valley and the surrounding interfluves.

Landscapes Formed by River Deposition

FLOODPLAINS

The valley of the Mississippi River, particularly in its lower course, provides many examples of land features associated with floodplains and deltas (Fig. 8-17). In its lower course the river flows through a valley ranging in width from 40 km to 200 km and over a floodplain with sediment several hundred metres deep near the mouth. Here the river is at base level, or very close to it, and the material it has been transporting is deposited. This deposition occurs in a number of places. Some is deposited in the bed of the river itself (for example, as a part of the meandering process, Fig. 8-9), while during floods, materials are deposited over much of the valley floor. The remaining material is carried to the mouth of the river where it is added to the enormous delta that is growing into the Gulf of Mexico or carried beyond the delta and deposited on the sea floor of the Gulf.

When the water level is low the floodplain is dominated by the wide and normally shallow Mississippi River. It often divides into numerous channels separated by constantly shifting sandbars. Because the lower Mississippi is very close to base level, it flows so slowly that its progress can be blocked by any minor obstacle in its course. The river may

then slightly alter its direction, initiating a meander (see Fig. 8-17). Eventually this meander will grow so large that the running water will cut through the neck of the meander to find a straighter course. This is especially common in times of flooding. Cutoff meanders usually remain as *oxbow lakes* for a short period, then slowly change to marsh, and finally to dry land. Meander scars may be seen all along the Mississippi and some of its larger tributaries. The development of meanders is a very rapid geological phenomenon. For example, the Rio Grande River, which forms the boundary between the United States and Mexico, has through its meandering caused boundary problems by 'transferring' territory from one country to another.

Levees are another type of landform produced by deposition in the bed of the river itself. Such deposition is greatest along the banks of the river where velocity is lowest, and a natural dike or levee is formed. As a result the entire level of the river is raised above the surrounding floodplain and the levees act like dams, keeping the flow of the river channelled. Levees on the lower Mississippi are 1 m to 3 m above the normal level of the water at the highest point (the bank of the river). Unfortunately, these natural dikes are seldom high enough to contain the river when the volume of the water is at its greatest. Usually a break in the levee will occur, and when this happens there is nothing to prevent the floodwaters from spreading over the adjoining floodplain (see Fig. 3-29). It is estimated that 140 000 km^2 of the Hwang Ho floodplain in China are liable to flooding in spite of attempts presently being made to control the volume of the river. Floods in the past have involved twice this area, and caused the death of many people through drowning or starvation.

Additional changes in landscapes are created because a river flows a little above its surrounding floodplain. The surface runoff of this slightly lower part of the floodplain can-

Fig. 8-14. The absence of trees in this photograph of the Missouri River near Portage, Montana, makes it possible to see the characteristics of a youthful landscape.

not be carried directly to the main river. This has often resulted in the development of many small tributaries called *Yazoo streams* (after the Yazoo River in the state of Mississippi). These streams flow parallel and close to the main river often for many kilometres—until elevations are suitable for a confluence. Much of the lower floodplain, however, remains poorly drained. Such areas, called *backswamps*, are very common.

DELTAS

Deltas are often extensions of floodplains. Normally it is difficult to distinguish where a floodplain ends and a delta begins. Deltas are formed by the deposition that occurs when the river joins a large body of water and its velocity is reduced. As the delta grows, the river divides into many channels called *distributaries*. These branch out in a fan-like pattern (Fig. 8-18). Usually only one or two distributaries are of any importance. The area of the delta is always being increased by the deposition of sediment at the mouths of the distributaries. Some distributaries may be important channels of the past that have diminished in size and now carry only local runoff.

Levees are formed along the banks of the distributaries as well as farther upstream on the floodplain. Since they are higher and better drained than the surrounding land, they provide the most suitable areas for human settlement. Because the remainder of the landscape is very flat, low-lying, and poorly drained, expensive drainage schemes and dikes to prevent flooding must be undertaken before the fertile sediment can be used for agriculture. (The gradient of the Nile delta, for example, is 1 in 10 000—on the average the river falls vertically 1 m every 10 000 m.) The need for drainage is particularly critical on the seaward margins of deltas. Here the line of demarcation between land and water is vague and changing. Severe storms

may inundate the lower parts of the delta on the coastlines unless they are suitably protected by dikes.

The size and shape of delta plains vary depending on the size of the river system, the amount of sediment carried to the river's mouth, and the strength of the tides and currents in the sea. Some of the more common types are shown in Fig. 8-18.

STUDY 8-7

1 Using a sheet of tracing paper, make a sketch of the area shown in Fig. 8-16. Locate the main river and some of the more important smaller ones. Also locate and explain each of the following features: meanders, oxbow lake, Yazoo tributary, floodplain, levees, backswamps, meander scars.

2 Make a simple cross section of Fig. 8-16 from the top to the bottom of the map passing through Smith Island and the Old and Red rivers. This cross section should show the shape of the main river channel and identify the area of maximum velocity, areas of deposition and erosion adjacent to the channel, as well as the features noted in question one.

3 Once a river meander is initiated it slowly develops from a small bend to a very large loop that eventually breaks through the neck of the meander, thus straightening the course of the river before new meanders are initiated. Referring to Fig. 8-16, make a series of simple sketches showing how this can happen. (Note that on the air photograph areas of recent deposition appear as white patches beside the river.)

Fig. 8-15. Weirgate Quandrangle, Texas-Louisiana (part of), United States Department of the Interior, Geological Survey. Scale 1:62 500. Contour interval 6 m.

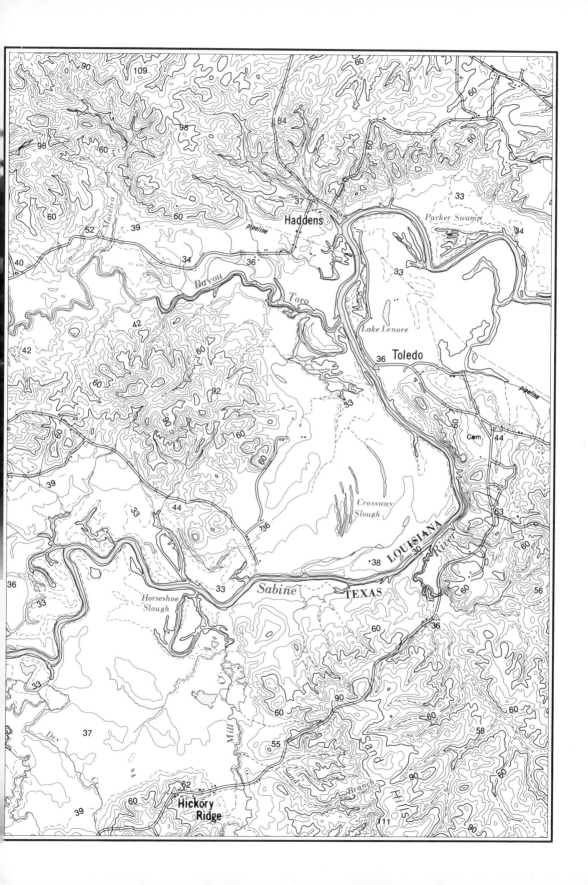

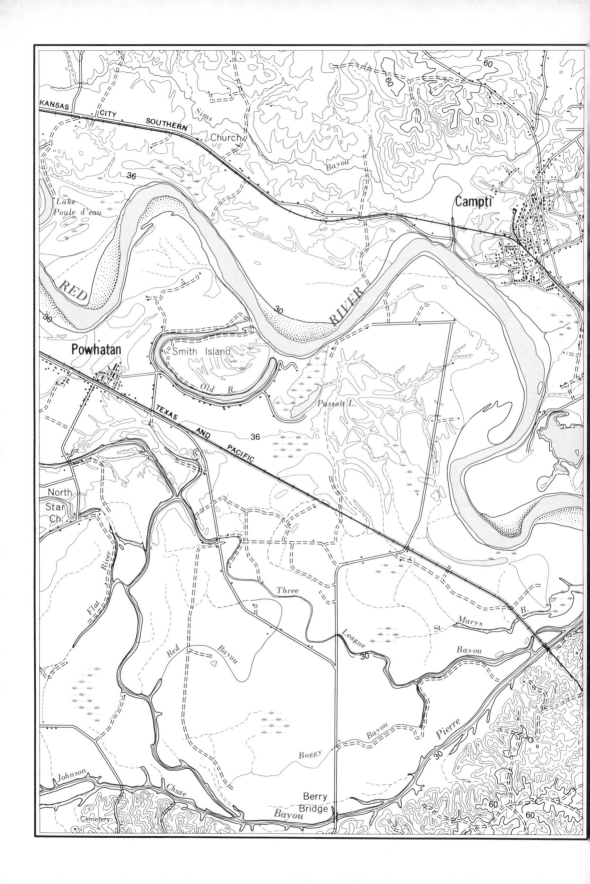

Fig. 8-16. Campti Quadrangle, Louisiana (part of), United States Department of the Interior, Geological Survey. Scale 1:62 500. Contour inverval 6 m.

Fig. 8-17. This vertical aerial view shows a part of the floodplain of the Red River, a tributary of the Mississippi, near Capti, Louisiana. Most of Fig. 8-20 is included in this photograph.

4 List the circumstances that are favourable to the formation of a delta and those that tend to prevent a delta from forming. Name some major rivers that do not have deltas and explain why this is so.

5 Describe and explain some of the variations in the size and shape of the deltas shown in Fig. 8-18. Give the reason for each name.

Karst Landscapes

Under normal conditions groundwater does not substantially alter the terrain. However in humid areas, where considerable thicknesses of soluble rocks such as limestone occur, the erosion caused by groundwater may become quite noticeable. This erosion is the result of a chemical reaction between ground water that is mildly acidic (carbonic acid) and soluble minerals in the rock such as calcium carbonate (see p. 193). Such erosion produces a type of landscape known as *Karst*. (The name was first used to describe the limestone country on either side of the port of Rijeka in Yugoslavia.) Rather than flowing on the surface, runoff seeps through fractures in the rock, slowly dissolving the soluble limestone and forming depressions known as *sinks*. These may eventually become tens of metres wide and five or more metres deep (see Fig. 8-19). Collapsed sinks occur where the bottom of a sink falls into caves that have developed beneath the surface. Such areas often consist of a maze of horizontal caverns and vertical shafts in which a variety of deposits can be found. These deposits include debris from ceiling collapse, sediment from underground drainage, and chemical deposits (largely calcium carbonate) that form dripstones or stalactites and stalagmites. Many well-known caves were formed in this way, including the Carlsbad Caverns in New Mexico and the Mammoth Caves in Kentucky.

Gradually an integrated drainage system develops below the surface, while on the surface only dry valleys and unconnected depressions remain. As erosion progresses the sinks and caverns grow larger and surface collapse becomes more common. Materials that have been dissolved or eroded are carried away by the underground drainage system. Slowly the whole land surface is lowered. The process of downcutting will continue in this way until the layers of soluble rock are removed or until a profile of equilibrium is established.

While there are many minor examples of the work of solution, there are only a relatively small number of locations in the world where the features described above occur over a large area. These include the Dalmatian coast of Yugoslavia, the Cumberland Plateau of Kentucky and Tennessee (see Fig. 8-19), southwestern China and northern Vietnam, and large areas in Florida and Puerto Rico.

STUDY 8-8

Figure 8-19 shows a part of the Mammoth Cave area of Kentucky. Using a stereoscope examine this photograph for evidence that shows it is a Karst landscape. How would you explain the difference between the northern and southern parts of the photograph?

The Work of Running Water in Sub-Humid Areas

In deserts and on their margins, where evaporation is normally greater than precipitation, running water is, surprisingly, still an important agent of land sculpture. Though there are few perennial rivers or streams, the runoff following the intense but infrequent rain storms may be very rapid and large volumes of water will move along normally dry valleys known as *wadis* or *arroyos*. Usually the land affected by the surface runoff is covered by large quantities of windblown and weathered rock fragments. This provides the overland and stream flow with abundant materials for transportation. Such streams, however, seldom last long. When the rain stops the runoff soon evaporates or sinks into the ground and integrated drainage systems seldom develop. Instead, the intermittent streams converge towards the nearest depression, depositing transported materials. Where sufficient water accumulates, intermittent lakes, known as *playas*, are formed. Examples of such basins of interior drainage (called *bolsons*) are shown in Fig. 8-20.

There are, however, some very important perennial rivers that flow across sub-humid areas and empty into the sea. These are known as *exotic rivers*. They originate in humid areas and receive enough runoff to balance the losses through evaporation that occur as they pass through the dry regions. The Nile and the Colorado are examples of exotic rivers.

The landforms of arid areas are normally quite different from those of humid areas, although similar gradational processes shape them both. Arid regions are characterized by steep, angular mountains separated by wide, alluvium-filled valleys intricately dissected by many dry river valleys. The steep and irregular landforms in arid landscapes contrast with the more subdued, rounded profiles of the hills and valleys in humid regions.

A cycle of gradation (see Fig. 8-25) has been recognized in arid regions that is similar to the cycle of gradation in humid areas. In its progression through this cycle the landscape becomes flatter as a result of the gradual wearing back of the hills or mountains and the filling in of depressions. In the early stages, streams that drain the hills or mountains adjoining the basins usually run through steep, gorge-like valleys until they reach the base of the hill or mountain, where the gradient changes rapidly. Most of their load is deposited here to form *alluvial fans*. Gradually these fans increase in size and eventually combine to form a continuous alluvial surface along the base of the mountain. This is known as a *bajada*. As erosion causes the mountain to retreat slowly, a flat erosional surface remains. This is called a *pediment*—rock surfaces usually covered by a thin veneer of alluvial material. At every stage in the arid cycle an intricate system of wadis dissects the surface of the basin. It converges on the lowest part of the basin where a flat, cracked surface of clay or salt—the dry remains of a playa lake—is often found.

The characteristics described above are found in many arid regions but are particularly common over large areas of the southwestern United States. They are typical of the landform known as plains with hills or mountains (see pp. 256-258).

Another type of landform develops in upland sub-humid areas underlain by horizontal sedimentary strata. Here rivers cut deep valleys, whose characteristics depend on the resistance of the rock strata that the river is cutting. The valley's sides may be formed alternately of scarps, gentle slopes, and shelves. Such a landscape is well known because of its spectacular development in the canyons of the Colorado River. An example can be seen in Fig. 9-3, which shows the plateau-like surface dissected by the Little Colorado River.

As erosion slowly progresses in upland sub-humid areas, residual hills protected by resistant rock are left dominating the landscape. The larger of these are called *mesas* and the smaller *buttes*. Where the surface layers are formed of weaker rocks such as clay or shale, erosion often produces a very rugged type of *badland* landscape with sharp-edged ridges and steep-walled gullies.

STUDY 8-9

1 In addition to surface water, what other

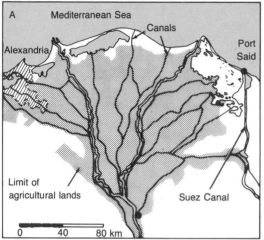

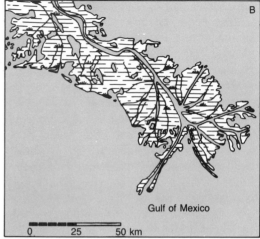

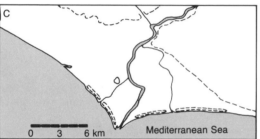

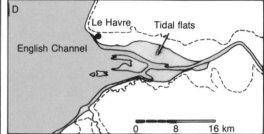

Fig. 8-18. Types of deltas: A the Nile, an arcuate delta; B the Mississippi, a bird's foot delta; C the Tiber, a cuspate delta; D the Seine, an estuarian delta (after Strahler).

gradational forces are responsible for the creation of landforms in sub-humid areas?

2 Compare each of the stages of surface-water gradation in an arid region with the equivalent stage in a humid region.

3 Figure 9-2 shows an area similar to that in Fig. 8-20. How were the hills in the foreground of Fig. 9-2 formed? What stage in gradation would these two areas represent? Identify the various features mentioned in this section.

Fig. 8-19. This stereo pair shows the Karst landscape in the vicinity of Mammoth Cave, Kentucky.

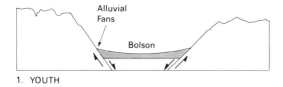

1. YOUTH

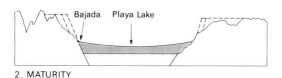

2. MATURITY

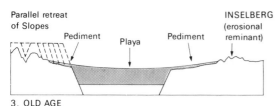

3. OLD AGE

Fig. 8-20. The cycle of erosion in an arid area.

GRADATION BY GLACIERS

Glaciers are formed in areas where temperatures are so low that snowfall does not melt entirely during the summer. Consequently, precipitation accumulates year after year as snow and later as ice. Such conditions are found today only at high latitudes or high altitudes. Glaciers found in high latitudes are called *continental glaciers* but are also known as ice caps or ice sheets; glaciers at high altitudes are called *alpine* or *valley glaciers.*

As snow accumulates it is compacted into *firn* or *névé*, a granular material denser than snow. As accumulation continues the firn is transformed by the weight of the overlying snow and firn into *glacier ice.* Under cold conditions, as in Antarctica, this transformation is a slow process. In areas where there is some melting in summer, this process will occur much faster. As the build-up continues,

the ice begins to move. The thicker continental glaciers flow outwards from the centres of accumulation, while the smaller alpine glaciers, aided by gravity, move downhill. The solid ice under pressure seems to possess sufficient plasticity to enable the whole mass to move forward at rates varying from a fraction of a centimetre to several metres a day, depending on the mass of the ice, the amount of slope, and the rate of snow accumulation.

The ice is thought to move by different mechanisms. In mountainous areas, where the pressure of the overlying ice causes the ice crystals to be deformed, a downslope movement takes place under the influence of gravity. This is known as *creep.* Glaciers also move by *basal slip*, which occurs when there is a thin film of water between the ice and the rock. Basal slip is particularly important under continental glaciers where temperatures in the ice rise with depth (tests near Byrd Station in Antarctica revealed surface temperatures of − 28°C rising to − 1.6°C at depths of 2 000 m). This is one of the principal ways in which erosion of the underlying bedrock occurs. A third form of movement is by *shearing* along planes in the ice, a process that is similar to faulting. Where ice in alpine or continental glaciers cannot adjust to tensional or compressional pressures produced by creep and basal slip, the ice shears and produces crevasses.

The velocity at which glaciers move varies not only from glacier to glacier but also within the same ice sheet. Measurements of a number of glaciers indicate an average movement of between 60 m and 150 m a year, although measurements of some fast-flowing Alaskan glaciers show movement up to 10 m a day, and movement as great as 24 m each day has been recorded on parts of the Greenland ice cap. The way in which an alpine glacier moves can be seen in Fig. 8-21.

The ice advances as long as the rate of snow accumulation is greater than the rate of

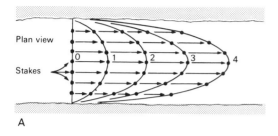

Plan view

Stakes

A

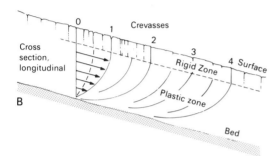

Cross section, longitudinal

B

Fig. 8-21. Relative flow-speeds within a glacier.

sheets were the fourth in a series of ice sheets that moved across approximately the same areas during the Pleistocene epoch. Between each of these advances there was an interglacial period during which the ice retreated. Although the last continental glaciers left southern Canada and northern Europe only about 11 000 years ago, it is not known whether we are in a fourth interglacial period or not. Reliable climatic observations were not made until the 18th century. The time since then is too short, geologically speaking, for conclusions to be drawn about temperature changes in the earth's atmosphere. We do not know if the atmosphere will continue to grow warmer, level off, or chill once more, thus causing glaciers to move south again (see discussion of climatic change beginning p. 109).

The reasons for the glacial movements of Pleistocene and earlier geological times have been debated for many years. There must certainly have been climatic changes involving either cooler summers or greater precipitation, or both. A number of theories have been developed to explain the ice ages.

Many hypotheses have been made to explain changes in the earth's climate that might have caused the ice ages (see also pp. 111 to 115). Some of these are based on known or supposed variations in the relation between the earth and the sun, particularly those involving variations in the earth's orbit and also variations in the inclination of the earth's axis. The principal deficiency of these

melting along the margins of the glacier. This melting process is known as *ablation*. When it equals the rate of advance, the glacier becomes stationary. If melting exceeds the rate of advance, or if the accumulation of snow at the centre of the glacier declines, the glacier will slowly melt back or retreat. Some glaciers terminate in the ocean, as, for example, many in Greenland and Antarctica. In these cases the glacier moves into the water and floats until large sections break away, forming icebergs and icefloes.

Past Glaciation

At the present time approximately 10 per cent of the earth's land surface is covered by glaciers. These are the remnants of much larger ice sheets that covered approximately 25 to 30 per cent of the earth's land surface about 20 000 years ago. These vast ice

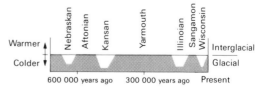

Fig. 8-22. Glacial and interglacial stages of the Pleistocene.

theories is that the ice ages thus produced would affect only one hemisphere at a time, but we know that the Pleistocene ice sheets were world-wide. In addition, climatic oscillations produced in this way would most likely be cyclical, that is, ice ages would occur on some kind of regular schedule every few hundred-thousand years. All evidence suggests, however, that this is not so (see Fig. 8-22). Ice ages are rare geological events.

Another set of hypotheses is related to the possibility that there have been variations in the amount of solar energy reaching the earth's atmosphere. George Simpson, a British meteorologist, advanced the hypothesis that the ice ages were a product of an increase in the amount of solar energy, as shown in curve A in Fig. 8-23. This in turn would cause an increase in temperatures (curve B), which would result in increased evaporation and, consequently, increased precipitation (curve C). Increased evaporation and greater cloud cover would keep air temperatures near the earth's surface lower than normal and lessen melting. In addition to increased precipitation, there would be increased snowfall and an accumulation of snow over the years. Glaciers would develop and remain in existence until temperatures became great enough to cause melting to exceed accumulation. This would result in the eventual retreat of glaciers, and an interglacial period. A new ice age would begin as the increase in solar energy declined and conditions were again suitable for ice accumulation. This second glacial period would end when solar radiation was inadequate to supply precipitation. According to this theory, one increase in solar energy (occurring over a long period of time) will produce two glacial stages. It follows, therefore, that there must have been two increases in solar radiation to produce the four ice advances that occurred during the Pleistocene. Unfortunately the principal weakness of this theory is that most evidence seems to point to a decline in tem-

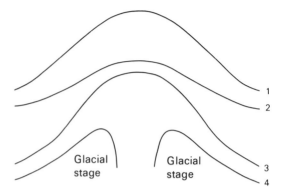

Glacial stage   Glacial stage

1
2
3
4

Fig. 8-23. This graph suggests how changes in solar-energy output would affect temperature, precipitation, and snow accumulation as shown by curves: 1) solar energy; 2) atmospheric temperature; 3) precipitation; 4) snow accumulation. Two such increases in solar radiation would be required to account for the four Pleistocene glacial stages.

peratures during the ice advances of the Pleistocene.

Another theory was put forward in 1966 by Maurice Ewing and William Donn. It is based on two suppositions: (1) the North and South Poles migrated to their present positions during the Miocene epoch, and (2) the Arctic Ocean alternated between ice-free and ice-covered periods during the ice ages (Pleistocene). According to this theory, the stage was set for the ice age to begin when the poles arrived in their present location. The actual development of glaciers occurred when the ice cover of the Arctic Ocean melted—an event that could be accomplished by a small change in the earth's temperature. As a result, the surface waters of the Arctic would warm up and there would be greater precipitation over land at high latitudes. Ultimately snow accumulation over the northern part of the continent would exceed melting and ice sheets would grow. As these glaciers became larger and larger they would cause temperatures in the northern hemisphere to drop. This would eventually lead to the freezing over of the Arctic Ocean so that it would no longer be a source of moisture. Glaciers extending far to the south would continue to exist, now fed by precipitation from the Atlantic and Pacific. Only when the surface temperatures of these oceans fell would evaporation from their surfaces decrease and the ice sheets begin to retreat. This process continued until the ice was almost completely gone. With the ice melted, atmospheric temperatures would begin to rise. Ocean temperatures rose more slowly than atmospheric ones so that increased precipitation from the warmer oceans did not begin until the ice had almost completely disappeared. (Ewing-Donn suggested the warming of the oceans lagged behind the atmosphere by 5000 years.) A new cycle of glaciation would begin when the Arctic became ice-free again, possibly as a result of the rise in ocean levels brought about by melting ice or changes in the earth's heat balance. The important premise of the Ewing-Donn theory is that changes in the amount of precipitation are the main controls of the ice ages and temperature changes play only a secondary role.

Another theory, advanced by Gilbert Plass, is based on the assumption that climatic variations sufficent to produce major glacial periods were caused by variations in the amount of carbon dioxide in the atmosphere. For example, an increase in CO_2 would cause temperatures in the lower atmosphere to increase because CO_2, together with water vapour and ozone, tend to prevent heat from escaping the earth's atmosphere. Plass suggested that the amount of CO_2 could have been reduced during periods of earth history when mountain building was particularly extensive. During these periods the exposure and weathering of igneous rocks would lead to the formation of carbonate rocks, a process that removes significant amounts of CO_2 from the atmosphere. Such periods of mountain building did in fact occur before both the Permian and Pleistocene. However, this theory does not clearly explain the various advances and retreats that occurred during the Pleistocene.

While no theory about glaciation can yet be validated, they are all important as a stimulus to discussion and the search for new evidence. On a more practical level, some scientists believe that changes we have caused in the atmosphere could eventually lead to a technologically produced ice age. For example, pollutants in the atmosphere have tended to reduce slightly the earth's surface temperature over the past several years. Whether such changes are capable of producing another ice age is not known, but most experts agree that only a relatively minor change in one of the links in the process of climatic change may be necessary to trigger other reactions.

STUDY 8-10

1 Why is it so difficult to establish an explanation of ice ages? Suggest the kind of evidence that might help settle some of the controversy.
2 What effect would the ice ages have had on plant and animal communities that existed during the Pleistocene? Of course humans existed during the last ice age; how would they have been affected?

Continental Glaciation

During the ice ages continental ice sheets differed from alpine glaciers. These differences can still be seen today in the ice caps of Antarctica and Greenland. Such ice sheets were much larger in area than alpine glaciers and their ice was much thicker. Today the Antarctic ice cap varies in area but at its maximum extent covers about 13 million square kilometres and is estimated to have an aver-

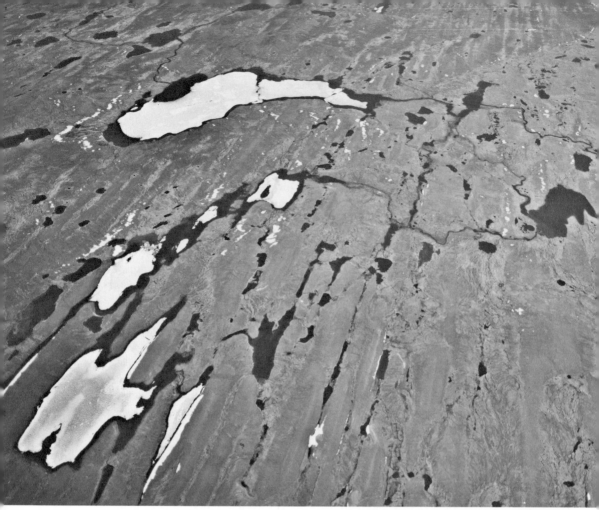

Fig. 8-24. This barren, ice-scoured landscape of the Canadian Shield in the District of Keewatin shows clearly the imprint of glaciation. Although much of this landscape is a product of ice scour, deposition features are also evident. The rock of the Shield is covered in parts by till, some of which has been deposited in the form of drumlins.

age depth of about 2000 m, reaching a maximum of over 4000 m in a few places. (It should be noted that the ice is thickest towards the centre, decreasing gradually outwards.) Because of the immense mass of the accumulating central ice, continental glaciers moved outwards in all directions from one or more centres and even moved up slopes.

THE WORK OF EROSION

Erosion under continental ice sheets occurred by abrasion. (It is still occurring in Antarctica and Greenland.) Rock fragments embedded in the ice act like a piece of sandpaper on the rock below the ice sheets, scratching and polishing it. Since temperatures rise as the depth of the ice increases, freezing and thawing of the ice cause underlying rock surfaces to shatter. This loose material is then quarried or plucked and incorporated in the ice sheet by the refreezing of meltwater.

Erosion was greatest under the central parts of the ice sheet. Towards its edges any erosion that may have taken place was usually marked by glacial deposits that occurred during the retreat of the glacier. The great continental glaciers that receded only a few thousand years ago reshaped large areas in both North America and Eurasia. The erosion and deposition accomplished by the glaciers had a general levelling effect, wearing down higher areas and covering much of the original landscape with thin deposits of material known as *glacial drift* or *till*. After the ice receded, river systems were re-established, so that the characteristics of the present terrain were formed by both ice and running water.

Features resulting from glacial erosion are most easily seen in areas of hard crystalline rock such as the Canadian Shield that covers almost half the area of Canada and a small part of the United States. These areas had only a thin pre-glacial mantle of weathered rock, and the hard underlying bedrock resisted the erosive action of the ice, yielding little new weathered material. Thus, when the glaciers melted there was little deposition to mask the results of glacial erosion. The surface of these plains is very irregular. There are many small differences in local relief with considerable variations both in the amount of sloping land and in the steepness of the slopes. While some places are almost level, in many others the local relief may be measured in hundreds of metres. The greatest irregularities are found in ice-scoured hills and mountains such as those in southern Quebec, the northwestern United States, and northwestern Labrador. Since these hills and mountains existed before the ice age, only their surface features resulted from glacial erosion.

Often in such landscapes the ice-scoured hills and gouged irregular valleys have a linear pattern (Fig. 8-24), with long narrow lakes and intervening ridges showing the direction in which the ice once moved. Parts of the surface consist of crystalline rock that has been polished and scoured and often bears scratches made by rocks carried in the ice.

These scratches are sometimes known as *striations*. The depressions in the landscape have only a thin covering of glacial drift and are of little use for agriculture. However, such depressions can support surprisingly dense stands of shallow-rooted conifers.

The drainage pattern in these areas has been greatly altered by the ice. Streams flow haphazardly with many falls and rapids along the glacially scoured valleys and form numerous lakes of various sizes and shapes. Over 10 per cent of the area of Finland consists of such lakes, while in many parts of the ice-scoured regions of Canada the percentage is even greater. These lakes occupy basins scraped out by the ice and valleys that have been blocked by irregular glacial deposits. Vegetation has taken root in many of the shallower lakes, converting them into swamps.

THE WORK OF DEPOSITION

The largest deposition of material transported by glaciers occurred, as might be expected, on the margins of the lands once covered by continental glaciers. In most instances this deposition made the landscape more level than it was in pre-glacial times. The extent of levelling, however, depended on the thickness of the deposits and the nature of the pre-glacial surface. In the several thousand years since the ice retreated, river systems have re-established themselves and considerably modified the landscape again.

Drift is the name given to all forms of rock debris deposited by glaciers as a result of melting. There are two main types of glacial drift: *stratified drift* and *till*. Stratified drift is made up of layers of sorted sands and gravels deposited by glacial meltwater to form such landforms as *outwash plains*, *eskers*, and *lake plains*. Till is an unsorted mixture of rock fragments of varying sizes left after the ice melted to form *rough moraines, ground moraines*, and *drumlins*. All of these types of glacial drift

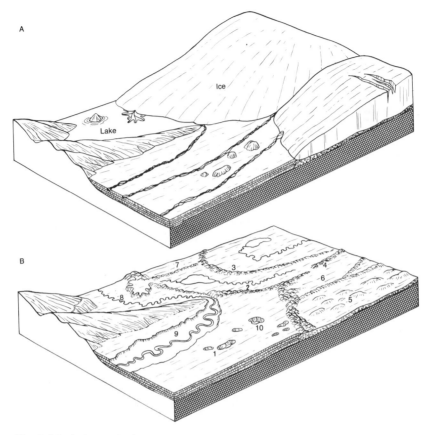

Fig. 8-25. A shows the farthest point of advance of a continental glacier and B the landforms that might be present in such an area today: 1) outwash plain; 2) terminal moraine; 3) recessional moraine; 4) interlobate moraine; 5) drumlins; 6) ground moraine; 7) esker; 8) lake plain; 9) spillway; 10) kettles.

are described below and many of them are illustrated in this section.

Rough moraines are formed where the glacier remained stationary for a period of time. They can usually be distinguished along the margin or in the middle of a ground moraine as patches of higher, stonier, and noticeably more irregular ground (Fig. 8-26). Their dimensions vary considerably, the larger ones forming conspicuous areas with numerous small hollows and hills, while the smaller ones are hardly distinguishable from the surrounding till plain. Deposits of sand and gravel, in places over 30 m deep, underlie the type of knob and basin relief that is characteristic of rough moraines. There are few streams on moraines, though larger rivers may cut through them, and there may be

Fig. 8-26. A variety of glacial landforms stand out in this high oblique air photograph, taken late in the day near Assiniboia, Saskatchewan. Identify as many of these features as possible.

occasional small lakes in the deeper basins. Surface water drains vertically through the permeable materials and re-appears as springs, which often indicate the edge of a moraine.

There are several types of rough moraine: *terminal moraines*, depositions that accumulated at the farthest extent of the glacier; *recessional moraines*, deposits that marked pauses during the retreat of the ice; and *interlobate moraines*, formed, as the name suggests, between two lobes or tongues of ice.

Ground moraine (or till plain) was formed from till deposits left by a glacier retreating at a fairly constant rate. These deposits are found covering large areas between the rough moraines. While seldom marked by any prominent landforms, except perhaps a rolling or fluted topography, the till may vary from deep deposits completely masking irregularities in the pre-glacial surface to thin, discontinuous deposits. Depositional features found on areas of ground moraine include clusters of *drumlins*, which are low, rounded

hills that look like half-buried eggs (Fig. 8-27), and *eskers*, that is, sinuous ridges of sand and gravel that were formerly the beds of streams that drained the front of the glacier through ice tunnels. Figure 8-33 shows the distribution of many of these depositional landforms for a small area of southern Ontario.

Outwash plains were formed from the debris washed out and carried away as the glacier melted. They are often conspicuous next to moraines because of their flatter surface. Small depressions in the ground (known as kettle holes) were formed when lumps of ice were left stranded after the glacier retreated. Like the rough moraines, outwash plains are formed mainly of sand and gravel, the finer silts and clays having been carried much farther from the glacier's front. On such plains agriculture is hampered by the coarse sand and gravel that often make the soil infertile and the water table very deep.

Much of the outwash material was carried away by glacial rivers. The deep, wide valleys these cut are known as *spillways*. This term is also applied to the drainage outlets that were developed when pre-glacial river systems were blocked by ice to form glacial lakes. Outwash material was deposited in much the same way as alluvium deposited by rivers. Most glacial stream beds are now drained by smaller streams (known as *misfit streams*) and have characteristics similar to those of outwash plains.

LAKE AND COASTAL PLAINS

Areas that have been glaciated abound in lakes. Some are a result of glacial scouring, others are formed where moranic materials have damned pre-glacial river channels, and many small lakes exist in kettle holes and other undrained depressions. Even larger lakes existed during glacial times when the receding ice acted as a dam, blocking the pre-glacial drainage outlet. For example, during the retreat of the Wisconsin ice sheet over 10 000 years ago, when parts of the Great Lakes were free of ice, the St. Lawrence River was still blocked. The waters of the lakes, therefore, rose to the next lowest outlet, which was first the Mississippi system and later the Hudson-Mohawk system. As a result the glacial Great Lakes covered a larger area than they do at present. For example, at one stage during the retreat of the ice, Lake Ontario (known to glaciologists as Lake Iroquois) had an altitude of 140 m above sea level compared with the present level of 74 m. All lakes formed in this way between higher ground and an ice front are known as glacial ponds or marginal glacial lakes. Some of the more important in Canada and the United States are shown on Fig. 8-28.

Where these lakes once existed, the present surface is formed of glacial sediments of clay and silt, or sand. Such lake (*lacustrine*) plains are among the flattest areas on the earth's surface, with meandering streams and large areas of marsh. Deltas were formed where streams flowed into these lakes or ponds. Today they can often be recognized as areas of sand deposits slightly raised above the surrounding land. These are known as *delta kames*. Lacustrine plains are also formed in areas where lakes once existed but under circumstances that have nothing to do with glaciation. For example, the Great Salt Lake in Utah is a remnant of a much larger lake (which geologists refer to as Lake Bonneville) that was formed when precipitation in this region of western North America was much heavier than it is today. Drainage flowing into Lake Bonneville resulted in even deposition over a wide area, and this became exposed as a flat plain when the waters receded to their present position.

The landform characteristics of coastal plains are a product of similar conditions undersea. Millions of years of deposition resulted in the formation of flat continental shelves. When parts of these are exposed a few metres above sea level—through a gentle warping of the earth's crust or through the lowering of the level of the sea itself—the land surfaces are generally very flat. The most prominent irregularities of these land surfaces are old wave-cut shorelines or cuestas caused by differential erosion (see Fig. 8-11). Such flat plains can be seen today in parts of the coastal areas of western Europe and the southeastern United States. In some cases, where the sea-bed was raised considerably above sea level, subsequent gradation by running water has destroyed the flat surface, carving an irregular plain instead.

STUDY 8-11

1 What effect has continental glaciation had on the terrain illustrated in Figs. 8-29 and 8-30? (Note that the Precambrian bedrock is close to the surface and outcrops in many places.) Is it possible to tell the direction in which the glacier moved?

2 How has the Wisconsin ice sheet affected the area shown on Figs. 8-31 and 8-32?

Fig. 8-27. A typical drumlin photographed east of Rochester, New York. What effect has this drumlin had on land use in the area?

Fig. 8-28. Glacial lakes and marine overlap in northern North America.

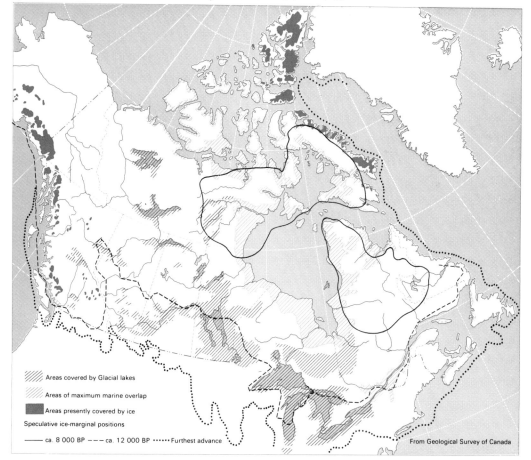

Areas covered by Glacial lakes

Areas of maximum marine overlap

Areas presently covered by ice

Speculative ice-marginal positions

—— ca. 8 000 BP — — ca. 12 000 BP •••••• Furthest advance

From Geological Survey of Canada

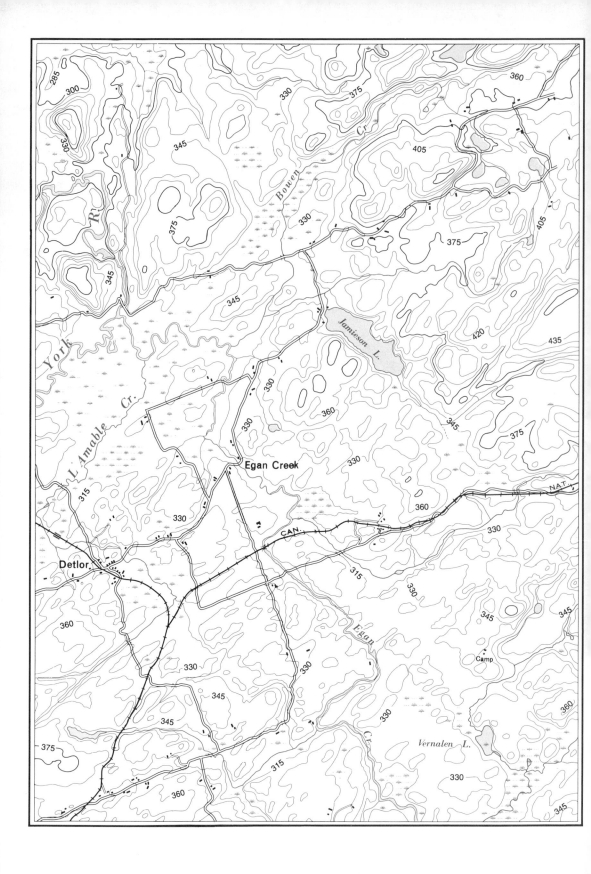

Fig. 8-30. Although this landscape has also been greatly influenced by the scouring action of glacial ice, much of the effect has been hidden by tree growth. Locate this area on Fig. 8-29.

Fig. 8-29. Bancroft, Ontario (part of), Canada National Topographic Series. Scale 1:50 000. Contour interval 15 m.

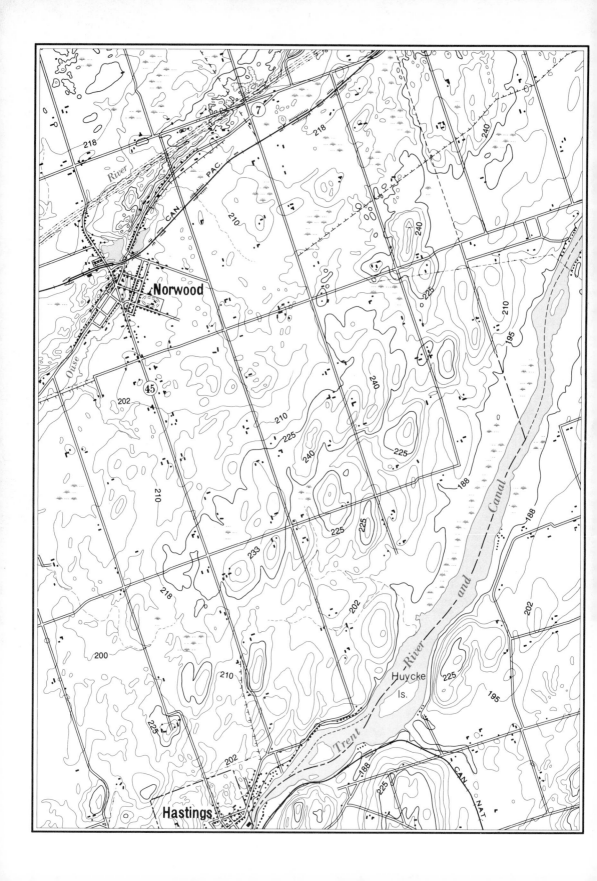

Fig. 8-32. Part of Fig. 8-31 is shown on this photograph. What has caused the slight jog in the east-west road? What accounts for the forested areas that have been left standing?

Fig. 8-31. Campbellford, Ontario (part of), Canadian National Topographic Series. Scale 1:50 000. Contour interval 7.5 m.

What was the direction of glacial movement? The feature parallel to Highway 7 out of Norwood (on the map) has been used by the Department of Highways as a source of gravel. What is this feature called and how was it formed?

3 For both areas referred to in the above questions, describe the use of the land, noting the distribution of, and the areas occupied by, forest, swamp, open water, agriculture, transportation, and settlement. Explain how these land uses may be related to the effects of glaciation.

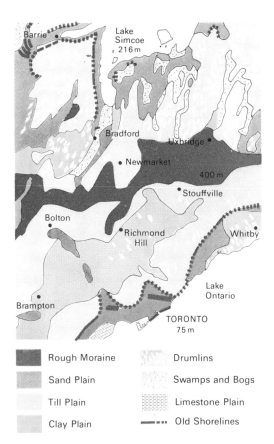

Fig. 8-33. Glacial landforms in a part of south central Ontario (after Chapman and Putnam).

4 A number of lake plains and coastal plains (areas of marine overlap) are shown on Fig. 8-28.

a How can you account for the large areas of marine overlap in northern Canada?

b Referring to an atlas to see the present surface drainage, explain how the glacial lakes were formed in northern Ontario (glacial lake Barlow-Ojibway) and Manitoba (glacial lake Agassiz).

5 The glacial landforms shown in Fig. 8-33 were produced by two lobes of the Wisconsin ice sheet; one occupied the Lake Ontario basin and retreated in an easterly direction and the other was located over the northern half of the map area and retreated in a northeasterly direction.

a Explain how the clay plains, old shorelines, rough moraine, and till plain were formed in relation to these retreating ice lobes. What type of rough moraine would this be?

b Where would you expect to find the most irregular and the most level landscapes in the map area? Explain.

c How might land use in the map area be affected by these deposits; that is, where would the best agricultural land be located and what areas would be best suited for recreational purposes?

Alpine Glaciers

Alpine glaciers, so called because they were first intensively studied in the readily accessible western Alps of Switzerland, originate in the mountain snowfields that form above the snowline. The altitude of permanent snow on any mountain depends mainly on the latitude but is also influenced by the amount of precipitation. Thus the larger glaciers originate from the accumulation of snow at high altitudes on the windward side of mountain ranges in middle and high latitudes.

During the ice ages alpine glaciers were considerably larger than they are today. In the middle and northern latitudes they covered areas far from the mountains in which they had their sources. Where they joined together on the level land below the mountains, they bore a distinct resemblance to continental glaciers.

Glacial ice produces the rugged, irregular features we associate with the major mountain systems of the world.

EROSION, TRANSPORTATION, AND DEPOSITION

Erosion by alpine glaciers occurs in different ways. At the head of a glacier and on its steeper slopes, cracks or crevasses in the ice permit meltwater (supplied by melted snow and surface layers of ice) to reach the rock underlying the glacier. This water penetrates the cracks in the rock and freezes, thus exerting tremendous pressure and shattering solid stone into small lumps or fragments. Rock pried loose by this process is frozen into the glacier and carried downhill. Embedded in the lower part of the ice, this debris acts as an abrasive; it gouges, scrapes, and polishes the bedrock over which the glacier passes. In addition to the rock quarried by the glacier itself, the ice picks up previously weathered rock fragments and either carries them in the ice or pushes them along in front of the glacier. The streaks in the glacier shown in Fig. 8-37 also show that a great deal of sediment falls or is washed on to the surface of the ice from the mountain sides above.

The result of this erosion, unlike that caused by running water, is a rugged and irregular landscape Fig. 8-34. One feature commonly found on upper mountain slopes or at the heads of valleys is a horse-shoe shaped, steep-sided depression called a cirque, (Figs. 8-34 and 8-35.) Cirques are so common a landscape feature in western Europe that there is a name for them in the language of every country where they are found, e.g., kar in German-speaking areas, cwm in Wales, corrie in Scotland, and so on. They were probably formed by intensive natural quarrying and ice scouring. (After ages of quarrying, a cirque may have a headwall several hundred metres high.) Very often after the ice has gone a small lake called a tarn will form in the basin of a cirque. If cirques are formed on either side of a ridge, the jagged, knifelike result is an arête (the French word for a fishbone or an edge). Where two of these cirques have broken through an arête from opposite sides, a gap or col has formed. Many cols are important mountain passes, such as those of Great St. Bernard, St. Gotthard, and Simplon in the European Alps. Where three or more cirques have grown together, a sharp pointed peak known as a horn is formed by the intersection of the arêtes. A classic example of such a peak is the Matterhorn in Switzerland.

In the lower part of a glacial valley, where the gradient decreases, the most important cause of erosion is the abrasive effect of the ice. Working on the bottom and lower walls of a V-shaped river valley from pre-glacial times, the ice deepens the valley. At the same time it steepens the valley sides and creates a curved, often U-shaped depression. The valley is also straightened as the glacier cuts off the ridges or spurs that the river wound around. In several parts of the world adjacent to the sea (for example, Scandinavia), glacial troughs were flooded by the sea as the glacier receded, and the sea level rose. These troughs now form narrow estuaries known as fjords. They are found mainly on west coasts bordered by mountains. Such coasts are characteristically irregular and indented, and can be easily recognized in an atlas.

Deposition from a glacier occurs for the most part during its retreat. As the ice melts at its lower end, the material that was embedded in it is released. (Considerable deposition may also take place at other times along the sides of the valley or between two

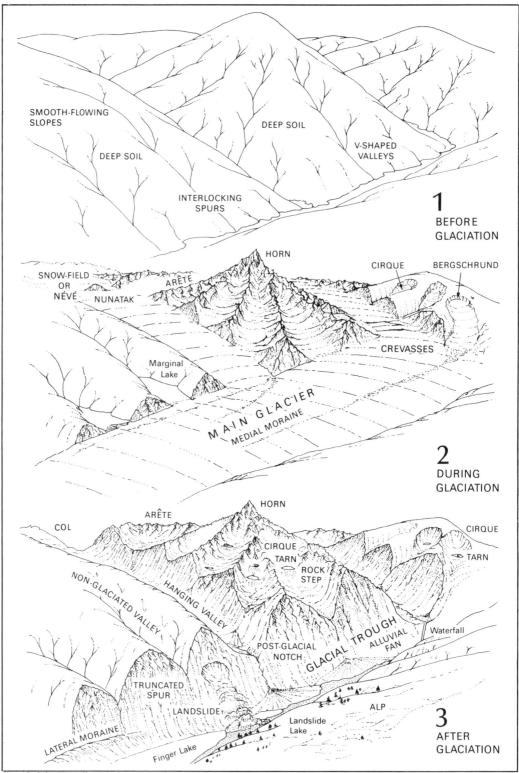

Fig. 8-34. The effects of alpine glaciation (after W.M. Davis).

Fig. 8-35. A series of small glaciers occupying cirques have formed an irregular arête in this portion of the Swiss Alps.

parallel glaciers.) Till deposits are left in much the same way as deposits from continental glaciers. They form several features: lateral moraines along the side of the valley; medial moraines between two glaciers; terminal moraines at the farthest extent of the glacier; and recessional moraines at various points during the retreat of the glacier. Most of these features are illustrated in Figs. 8-36 and 8-37.

The fine material carried by glacial meltwater accumulates in deposits of stratified drift along the valley floor. The heavier material—deposited first—is known as *valley train*. There are many cases where these or morainic deposits block whole valleys, impounding streams and forming small lakes. Picturesque lakes such as Lake Como in Italy

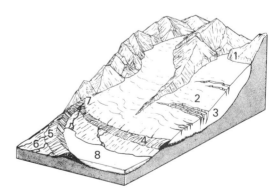

Fig. 8-36. Alpine glaciers and moraines: 1) headwall of a cirque; 2) crevasses; 3) eroded material embedded in the ice; 4) recessional moraine; 5) terminal moraine; 6) valley train; 7) lateral moraine; 8) moraine-dammed lake.

Fig. 8-37. The retreat of these glaciers has left a rock-strewn mountain slope with sharply defined marginal and medial moraines.

and Lake Louise in Banff National Park, Alberta, have been formed in this way.

STUDY 8-12

1 a Using the identifying points A-B and C-D in Fig. 8-38, make two cross sections of the valley between these points. Identify and describe the characteristics of the valley illustrated by the two sections.
b Using a sheet of tracing paper or a photocopy of the contour map (Fig. 8-38), identify examples of each of the following: cirques, arêtes, U-shaped valleys, tarns, hanging valleys, rock steps, finger lakes, cols, horns, and areas presently covered by ice. Briefly explain how each was formed. Many of these features can be found more easily on the map if they are examined first on the stereo pair of air photographs (Fig. 8-39) through a stereoscope.
c Examine the location of existing snow and ice fields and explain their present restricted locations.
2 Using an atlas, locate some of the major areas in the world where alpine glaciers exist today. The altitude at which glaciers are now found varies from place to place. Why does this variation occur? What is the extent of the variation?
3 Comment on the differences between recently glaciated mountains and mountains and hills where glaciation has not occurred or where glacial features have been obliterated.
4 Referring to the various illustrations in this section, describe some of the more important characteristics of (a) streams and (b) examples of mass wasting in mountainous regions. Comment on the general importance of these gradational processes in mountainous regions.

GRADATION BY THE WIND

Compared to the other agents of gradation,
in most regions wind plays a minor role. However, it can have a considerable effect over limited areas where there is little precipitation and consequently little vegetation to hold surface particles together. Gradation by wind is confined, therefore, mainly to arid areas.

In humid regions a significant amount of wind erosion occurs in areas of beach sand, during periods when river beds have dried up, or when loose top soil is unprotected in cultivated fields. People have unwittingly assisted wind erosion by ploughing agricultural land where the precipitation is low and unreliable. Huge quantities of valuable top soil have, as a result, been blown away in the drier parts of dustbowls of the United States, the USSR, and Canada.

In times past, wind erosion on a large scale occurred after the retreat of the continental glaciers. Loose deposits unprotected by vegetation were picked up and carried great distances in dust storms.

EROSION, TRANSPORTATION, AND DEPOSITION

Wind is able to pick up small particles and even to carry them for considerable distances. This process, known as *deflation*, occurs where there are few obstacles to impede the wind. While deflation is influenced by the configuration of the landscape, it is not affected by the landscape in the same way that running water is. The effect of deflation varies, but usually the surface is not lowered to any great extent nor are any unusual landforms created. The exceptions are occasional shallow depressions known as *blowouts*. The lower limit or baselevel for wind erosion is fixed by the water table. Because the wind can only carry small particles, a landscape affected by deflation may consist largely of a ground covering of gravel, coarse pebbles, and larger fragments of rock.

Much of the surface of the world's deserts

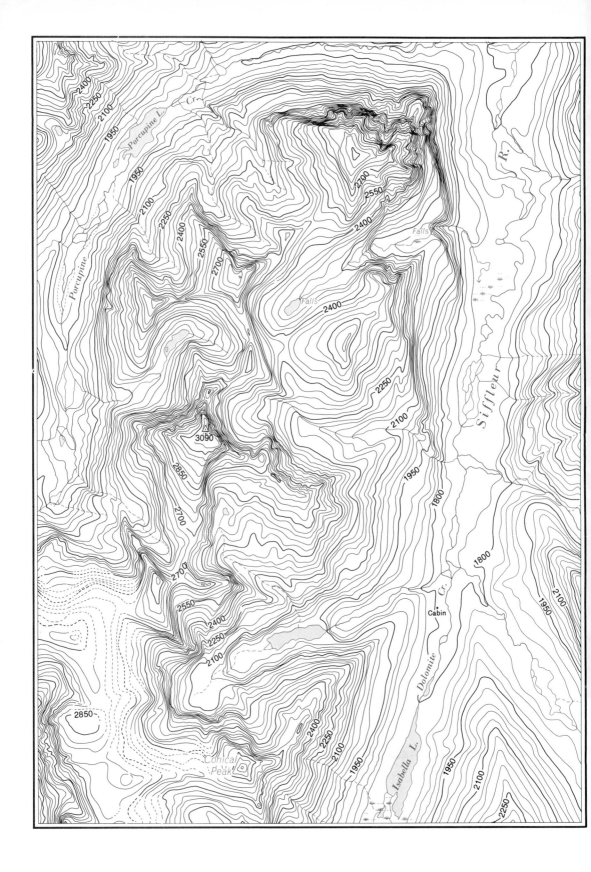

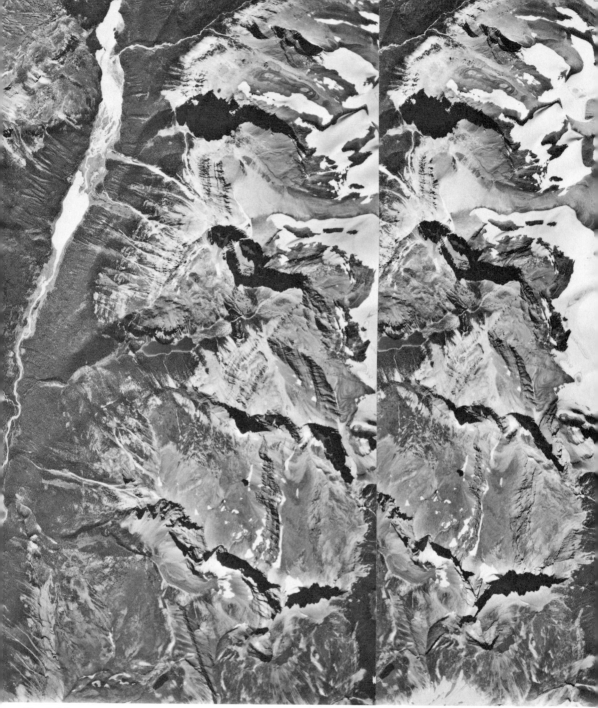

Fig. 8-39. Part of Fig. 38 is shown on this stereo pair.

Fig. 8-38. Siffleur River, Alberta (part of), Canadian National Topographic Series. Scale 1:50 000. Contour interval 30 m.

consists of material left behind after the wind has carried away finer particles. The surface ranges from coarse sand and gravel to large stones and rocks. It is sometimes known as *desert pavement* or, when it consists mainly of gravel, as *reg*.

Some of the wind-modified features of irregular plains in desert areas are described in the passages below:

The desert hollows [of the Gobi Desert] *could not possibly be excavated without the work of wind; no other agency could lift material out of an enclosed lowland. During violent sandstorms the air was dark with flying dust and the sun was dim and red. Much of the stirred-up sand and dust settles upon the surrounding country, but no doubt some of the finer material is exported to great distances, even beyond the rim of the Gobi. The sides of the hollow are dissected by rainwash and by rills from the upland. During epochs of active erosion, the sloping bluffs are fretted into typical badlands, and the loose sediment is washed down to the floor of the hollow. As soon as the thin flat apron of sediment is dry, it is exposed to the tireless winds which carry much of it out of the hollow.*[2]

The stones of these plains [the Stony Desert, east of Lake Eyre in Australia] *are angular, and are said in places to fit together with the accuracy of a mosaic; and when the pebbles are thus closely packed, patches of the Stony Desert appear like a tesselated pavement. The Stony Desert is due to the absence of water. The country where it occurs was once covered by a sheet of the rock known as desert standstone, in which there are abundant pebbles of quartz, sandstone and other hard materials. The desert sandstone has slowly decayed under the action of the weather; the loose sand has been blown away by the wind, and the hard fragments remain scattered over the ground.*[3]

Particles transported by the wind act as abrasives and loosen other rock fragments.

Though this sandblast effect is widespread, it is most noticeable on the lower parts of sloping surfaces since coarser particles seldom get more than a few metres off the ground. In areas where wind abrasion has been particularly strong, unusual features resembling pedestals, windows, and arches are found.

Fine particles deposited by the wind are found on almost all parts of the earth's surface. There are only a few areas, however, where the deposits occur in any depth, adding to sand deserts in arid regions. In more humid areas such deposits form a fine-grained material known as *loess* (from the German *löss*, meaning a fine, yellow-grey loam). Similar areas of sand that have been moved a short distance by the wind are found next to large beaches. On exposed coastlines sand is quite often moved several kilometres inland with the result that beaches may be backed by a strip of sand dunes of variable width. In such areas vegetation exists rather precariously and if people upset the balance the sand may begin to move. Fortunately, methods to re-establish vegetation have been developed—pine trees are often the best means of stabilizing the sand.

Very fine material can be transported great distances by wind, but most particles are deposited only a short distance from their place of origin. Deposition occurs when the velocity of the wind is reduced or the amount of moisture in the air increases. Most of the deposition occurs in arid areas as a result of a reduction in wind velocity. Where the deposition is substantial, sand deserts (referred to as *erg*) are the result. The landscape of these deserts makes up approximately 15-20 per cent of the area of the world's deserts. Along with other areas of sand deposition, such as beaches in humid climate, these deserts are almost always characterized by wave-like hills or ridges of sand called *dunes* (Fig. 8-40). It should be noted, however, that dunes are not just confined to sand deserts but are often found in deserts over barren bedrock. What-

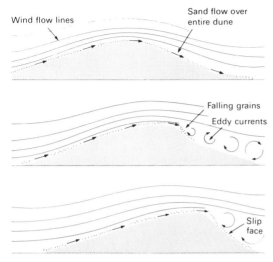

Fig. 8-40. Evolution of a sand dune.

ever their location, dunes are usually formed around a small obstacle that breaks the even flow of the wind and causes the sand to build up (see Fig. 8-41). Eventually they become large enough for their size to interfere with the wind, causing further growth. Most dunes gradually shift their position. Their rate of movement as well as their size and shape vary considerably from place to place, depending on the strength of the wind, the amount of material deposited, and the amount of vegetation present.

Material eroded by wind and carried beyond the desert is often deposited in considerable quantities on humid margins of dry lands. These loess deposits can be found in small quantities over a large part of the earth. Deep deposits are usually confined to the leeward side of major deserts and areas adjacent to the deposits of fine materials left by continental glaciers. Substantial loess deposits were also formed over areas of North America and Eurasia during glacial times when strong winds off the ice sheet carried fine rock particles in a southerly direction.

Some of the best examples of loess deposits are found in northern China, central Europe, and the central United States. In these places, the landscape has steep valley slopes with flat land between them as a result of the vertical, fine-grained structure and porous nature of loess. Such flatlands are marked by gullying, particularly where the vegetation cover is thin. Slumping is also common, giving the valleys a step-like profile.

STUDY 8-13

1 In general terms, explain how the gradational work of wind differs from that of running water.
2 Even in desert areas the gradational action of running water is normally more important than that of wind. Why is this so?
3 In sub-humid areas the work of all the gradational agents in lowering the landscape proceeds much more slowly than in humid areas. Explain the reasons for this.
4 Explain why some areas of desert are covered by sand and others by particles of rock and gravel.
5 Examine Figs. 8-41 and 8-42 and describe and explain how different types of sand dunes are formed. What are some of the problems that result from this shifting sand?

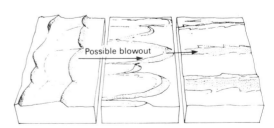

Fig. 8-41. Different types of dunes: transverse to the left, parabolic in the centre, and longitudinal to the right. The arrows indicate the way in which one might change to the other.

Fig. 8-42. These crescent shaped dunes, known as barchans, were photographed in Death Valley, California. Some idea of their size can be judged from the two people standing near the middle of the photograph.

6 Referring back to pp. 193 to 194, explain why mechanical weathering is more important than chemical weathering in dry areas.

GRADATION BY WAVES AND CURRENTS

Over the centuries the ceaseless pounding of waves has produced a variety of different landforms along the world's coastlines. The factors involved in their development are very complex and no simple classification of shorelines exists.

Although river and glacial erosion and deposition may have had considerable influence on the development of certain coastlines, the most important factors are waves and currents. The work of these agents is complicated by the fact that the levels of land and sea—relative to each other—have undergone many changes even in recent geological times. Figure 8-43 shows the changes in sea level that have occurred over the past 17 000 years. The sea level has been rising over this period, gradually submerging coastlines. There are also many areas where crustal movement has occurred, either raising or lowering the land in relation to sea level. In most cases the present shoreline of any coast has been established in fairly recent geological times.

Simplifying the great variety of different types, three kinds of shoreline are described: embayed, beach and bar, and cliff and terrace.

EMBAYED SHORELINES

Embayed shorelines, commonly known as drowned or submerged coastlines, are indented and irregular because they were formed by the drowning or submergence of valleys. Where the valleys were originally formed by rivers, the term *ria* coast is used; where the valleys were formed by alpine glaciers, the term *fjord* coast is used. The characteristics of these shorelines were largely determined by the type of landscape that existed before the land was submerged, and there are many possible variations. All of these shorelines, however, are highly irregular.

The outline of every kind of shoreline is constantly being modified. On embayed shorelines, the greatest force of the waves is concentrated on the headlands (Fig. 8-44),

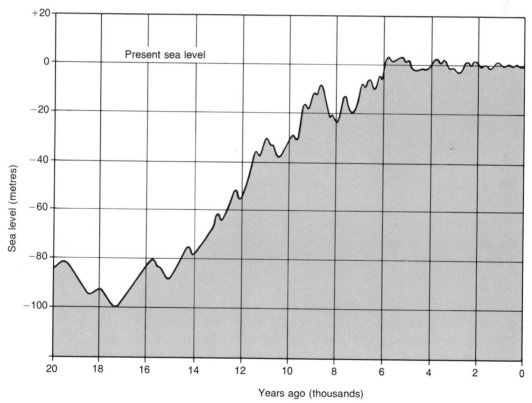

Fig. 8-43. The oceans reached their lowest level approximately 17 000 years ago, at the height of the last glacial period (Wisconsin). Sea level has remained relatively constant for the past 6000 years, and even the amplitude of the short-term variations has been decreasing (after *Scientific American*, Vol. 202, No. 5, May 1960).

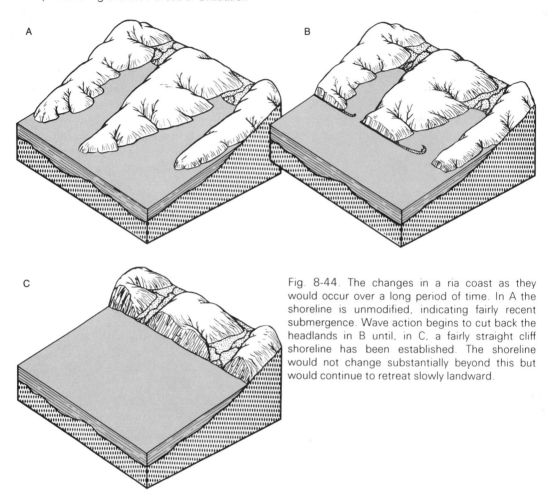

Fig. 8-44. The changes in a ria coast as they would occur over a long period of time. In A the shoreline is unmodified, indicating fairly recent submergence. Wave action begins to cut back the headlands in B until, in C, a fairly straight cliff shoreline has been established. The shoreline would not change substantially beyond this but would continue to retreat slowly landward.

which are gradually being worn back. At the same time river deposition is filling in the estuaries. Ultimately a relatively straight shoreline will result. The comparatively short period of time that has elapsed since land and ocean reached their present levels explains why so few straight shorelines exist.

BEACH AND BAR SHORELINES

Beaches along the shore and sand bars off the shore are common phenomena on many coastlines. They are found where the water off a shoreline is shallow for some distance out to sea and the waves are relatively constant in direction and size. Beaches occur as continuous strips on low, smooth coasts. On the more irregular coasts, they are found only in bays and inlets separated from each other by rocky points and promontories.

Off-shore bars (Fig. 8-45) such as those found along much of the eastern and southern shorelines of the United States were probably formed by waves breaking in shallow water some distance from the shorelines. The scouring action of the breaking waves heaps up sand until it is visible at low tide. (Often a bar continues to grow until it forms a barrier,

Fig. 8-45. A typical beach and bar shoreline on the east coast of New Brunswick (altitude 3550 m, lens focal length 150 mm).

even at high tide.) Separating the bar from the shoreline is a lagoon, which in many instances is quite shallow and bordered on the landward side by swampy tidal flats. A lagoon of this type tends to be a temporary land feature because the sand bar is being gradually moved landward by the action of the waves, particularly during storms. Ulti-

mately the sand will be driven on shore where it may form coastal dunes.

CLIFF AND TERRACE SHORELINES

Most embayed shorelines and beaches and bar shorelines are more or less temporary features that have resulted from changes in the

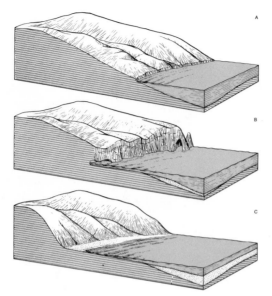

Fig. 8-46. The development of sea cliffs usually occurs where land slopes steeply into the water. Wave action gradually carves out a small cliff (a nip) as shown in A. This action continues to enlarge the cliff, mainly through undercutting, as is illustrated by the wave-cut notch at the cliff base in B. (Unusual features such as sea caves and sea arches are also often found on this type of shoreline.) As the process continues, a sloping rock platform is built under the water adjacent to the cliff. Finally this platform becomes so wide that wave action decreases, a beach develops, and the cliffs become more subdued.

relative levels of the land and water. There are many shorelines, however, where waves breaking directly on the shore have produced a fairly regular coastline with distinctive cliffs and terraces. These are normally found along rocky shores where the land rises steeply and the water deepens rapidly out to sea. As illustrated in Fig. 8-46, wave action has undercut and driven back the shoreline, producing a cliff. The area cut away in front of the cliff— together with material eroded from it and deposited by the waves—forms a terrace. Where the relative levels of land and sea have

changed, a series of cliffs and terraces may be left. These are submerged beneath the sea or stand high and dry as part of the coastal landscape.

STUDY 8-14

1 Referring to Figs. 8-44, 8-45, and 8-46, sketch a simple map and a cross section showing the general characteristics of each type of shoreline; add notes explaining the formation of each shoreline.

2 Referring to an atlas, state the general type of shoreline found in the following areas: southern England, Texas, the Red Sea, western Ireland, the Netherlands, North Carolina, and southern Chile.

3 The characteristics of shorelines are important in the development of harbours for use by large ocean-going ships.

a Comment on the physical suitability for the development of harbours of each of the three types of coastlines discussed.

b The value of a particular location for a harbour depends more on the need for a harbour at that location than on the physical characteristics of the coastline. Discuss this statement.

c Look up in an atlas each of the port cities noted below. On what type of coastline is each located? Which ports have almost 'natural' harbours, and which depend on considerable construction and maintenance? New York City, Callao (Peru), Rotterdam, Sydney, Rio de Janeiro, San Francisco, Los Angeles, Vancouver, and Genoa.

[1]M. Morisawa, *Streams, Their Dynamics and Morphology* (New York: McGraw-Hill Book Company, 1968), pp. 73-4.

[2]C.P. Berkey and F.K. Morris. *The Geology of Mongolia* (New York: The American Museum of Natural History), 1927.

[3]J.W. Gregory. *The Dead Heart of Australia* (London: John Murray Ltd.), 1906.

9/ TYPES OF LANDFORMS

Up to now we have examined various processes responsible for creating the earth's landforms. Theories about how landforms came into being have been revised and modified as new knowledge has come to light in the past, and further revisions are likely to be necessary in the future.

In this chapter we are going to look at the classification and description of landforms. Enormous variations in landscape can be observed from place to place and these must be classified if the landform characteristics of any area are to be described. The first distinction is between five major landscape types: plains, mountains, hills, plateaus, and plains with hills or mountains. Each of the main types may then be subdivided into various kinds of plains, hills, and mountains. The world distribution of major types of landforms is shown on Fig. 9-1.

The terms *plain, hill, mountain,* and *plateau* each have a commonly understood meaning. However, it is necessary to define the terms more precisely in order to use them as the basis of a classification of landforms. It is fairly easy to distinguish different types of climate because elements common to all types—such as temperature and precipita-

tion—are quite specific and capable of precise measurement. Landforms, on the other hand, have only a small number of easily distinguished or measurable characteristics. To complicate matters further, there often seems to be no apparent pattern to the differences in landforms from place to place. It might even be said that no two areas of land have exactly the same detailed landform characteristics. Because it is the task of the geographer to describe and interpret the landform characteristics of different areas, some method is necessary to permit a systematic description of different areas with widely varying terrain.

At this point it should be noted that the analysis of landforms is generally carried out from topographic or contour maps. Most of the analysis on the following pages is based on these maps. Air photographs (particularly stereographic pairs) are also of great assistance in analysing landforms and, of course, field work (where practical) is important. Many of the techniques used in this landform analysis are explained in the Appendix on Maps and Mapwork. The section on contours should be thoroughly understood before proceeding with the material that follows.

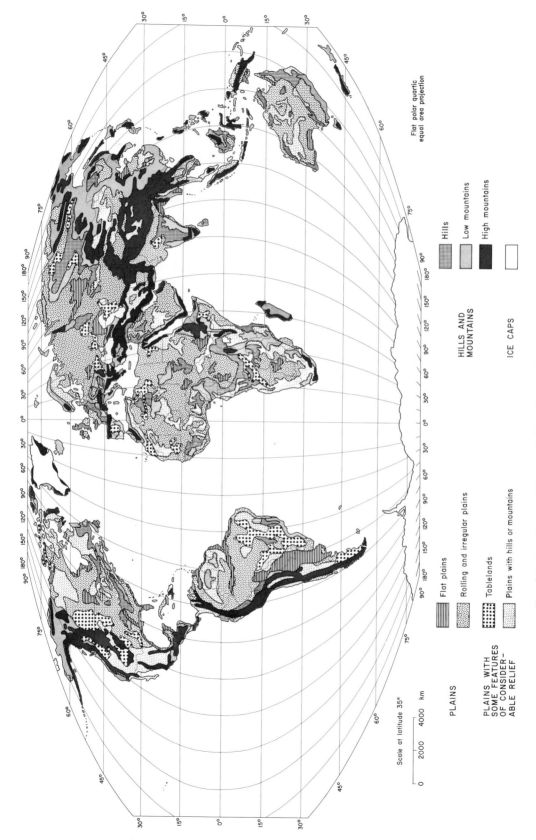

Fig. 9-1. Landforms of the earth (after Trewartha. Robinson. and Hammond).

PLAINS

| Flat plains |
| Rolling and irregular plains |
| Tablelands |
| Plains with hills or mountains |

PLAINS WITH SOME FEATURES OF CONSIDERABLE RELIEF

HILLS AND MOUNTAINS

| Hills |
| Low mountains |
| High mountains |

ICE CAPS

Flat polar quartic
equal area projection

Scale at latitude 35°

km

0   2000   4000

If many small areas are carefully compared in order to determine precisely how the landforms of each is unlike that of the others, a long list of specific differences is soon accumulated. And if this list is analyzed, it becomes apparent that many differences may be grouped under the four major headings of slope, surface material, arrangement, and dimensions. That is, the differences between any two sections of the land surface may be expressed in terms of these four major topics.[1]

Slope

Slope is defined as the number of degrees the surface of the ground is inclined from the horizontal. In this book the different categories of slope are defined in the following way: *gentle* is used to describe any slope under 4° from the horizontal; *moderate* for any slope from 4° to 10°; and *steep* for any slope over 10°. The method of calculating slope is described on p. 293. The different areas of sloping land in any single region can be shown on a single-topic map. Such maps, described on p. 295, are intended to give information on only one topic, e.g. different areas of sloping land, the drainage pattern, etc.

Within any area there may well be several different kinds of slopes. However one type is likely to predominate. For example, in mountainous regions the largest part of the land surface will consist of steeply rising land even though some parts of the land—particularly in the valley—may be gently sloping, and other parts will have moderate slopes.

Surface Material

When analysing a landscape it is important to take into account the nature of the surface material, particularly when it is something other than soil. The major types of surface material are soil, regolith, bedrock, swamp, open water, and ice. Contour or topographic maps will be of only limited assistance in distinguishing one type of surface material from another. Aerial photographs are the best source of information if direct observation of the area is not possible. A single-topic map can show where considerable variation exists in an area.

Dimensions

Any description of a landscape, whether by words or maps, should include an indication of the dimensions of its features. Dimension depends on measurements of vertical differences (i.e. variations in elevation or height) and measurements on a horizontal plane (i.e. variations in width or area).

Vertical differences can be expressed either as elevation above sea level or as local relief. The term *local relief* refers to the difference in elevation between the highest and lowest points in a specific area, that is, it is used to indicate the height of ridges, hilltops, or upland areas above adjacent valley floors.

Profiles indicate how a terrain appears in cross section. They are a useful device to show not only differences in local relief but also differences in the amount of land in the three slope categories. Profiles are helpful in visualizing the terrain when the arrangement of contour lines on a map is confusing. One or two profiles are usually sufficient for an overall impression, however more may be necessary for an area that has complex landforms to show the shape of such features as ridges, escarpments, and valleys, or to show the gradients along rivers.

Horizontal dimensions indicate the spacing and the area of different features. Sufficient measurements should be made to establish the distance between rivers, the width of valleys, the distance between peaks or ridges, the extent of swamps and tracts of gently sloping or steeply sloping land, and so on.

Arrangement

We have just examined three characteristics of landforms. We now need to develop a general method of describing the arrangement of these characteristics in any area.

If the characteristics of the entire area are fairly uniform, or if the landscape is a simple one (such as a flat or gently rolling plain), the description will consist of a brief summary of the essential features. In many cases, however, there will be great variations from one part of the area to another. For example, one section may have very gently sloping land with shallow river valleys and many areas of poor drainage; another may have an irregular surface with several low hills, many small differences in local relief, substantial areas of moderately sloping land, and deep v-shaped valleys; and a third may consist of the steeply sloping face of an escarpment. (These are only three of an unlimited range of possibilities.) When such variations exist it is necessary to show them on a single-topic map, describe the features of each section, and indicate the relationship of one section to another.

Explanation

After completing the description it is usually possible to interpret many of the tectonic and gradational processes responsible for the formation of features of the landscape. However, it is not always easy to be sure about the forces that created all the features of a landscape. If in doubt, list a number of possible explanations. Often when a number of alternatives are given, one of them appears to be more logical than the others.

STUDY 9-1

1 Sketch a single-topic map to show the different categories of slope in the area described by Fig. 8-12. Estimate as best you can the area of land in each of the three categories.

2 Sketch a single-topic map showing the different types of surface material in the area described by Fig. 8-12. It can be assumed that the steeper slopes have a thin covering of regolith and the very steepest areas show bedrock at the surface.

3 a Describe the principal differences in local relief in the area shown in Fig. 8-12.

b Prepare one or two profiles of Fig. 8-12 from west to east. In addition, make several small profiles across different parts of the valley, and also make a profile indicating the gradient of either Brooms Brook or Ryans Brook.

c What is the area of the map? From a reasonable number of measurements estimate the average width of each valley and each interfluve.

4 What patterns can be seen in the arrangement of the landforms in the area on Fig. 8-12. Summarize the principal features of this particular landscape, indicating how the area might be divided on a single-topic map into several sections, each with its own distinctive characteristics.

5 Explain as best you can the forces responsible for the evolution of the landscape shown on the map.

DEFINITION OF TYPES OF LANDFORMS

It is possible to make fairly precise definitions of the four types of landforms based on slope, surface material, and dimension. Definitions do not involve arrangement of terrain because so many variations in arrangement can occur within each of the major types of landforms.

Plains

The distinguishing characteristics of plains are low local relief and flat to gently sloping surfaces. Since the amount of local relief can vary as much as 80-100 m, it is incorrect to stress the level character of plains.

It is estimated that plains make up approximately 55 per cent of the earth's land surface. They can have originated in a number of ways. One type of plain known as a *displain* or *discordant plain*, occurs when tectonically produced structures are levelled by erosion. In such a situation there is a marked difference between the fairly level surface produced by the gradational forces and the complex underlying rock structure that was altered by folding and faulting. Many shield areas are examples of such plains. Notable examples of these are the Baltic and Canadian Shields and the Piedmont Belt of the Appalachians.

The second type of plain is known as a *conplain* or *concordant plain*. It is characterized by a conformity between the surface of the plain and the underlying sedimentary strata, which may be either horizontal or gently warped. The surface materials of these plains were once deposited under ancient seas. After being exposed as land, they were shaped by gradational agents. Some plains are very level and have surface features that are mainly a product of river-deposited materials. The coastal plain of the Gulf of Mexico is an example. Other plains have been raised or gently warped. As a result gradational forces have cut into underlying strata and produced deeper valleys, escarpments, and in general a much rougher type of surface. An example of this is the plain of the Paris Basin (Fig. 7-16).

Gentle slopes and an absence of high local relief exist in areas where tectonic activity is sufficiently remote in geological time for the gradational forces to have worn down or covered any features of high relief. Such gentle slopes are also found where tectonic activity has occurred so recently that streams have not had time to dissect the uplifted surface. (Several of these situations may, of course, be combined in any specific area.)

As indicated on Fig. 9-1, plains can be subdivided into two types: flat and rolling or irregular.

FLAT PLAINS

Most flat plains were formed by deposition from streams and rivers or by deposition that occurred under lakes and seas. In the latter case, the lakes have disappeared or the sea bed has been raised. Such plains have very small differences in local relief and extremely gentle slopes. This is a result of two factors: first, the type of deposition that formed them; and, second, the fact that the surface of these deposits is seldom more than a few metres above the present base level. A good example of the various features that may be found on a plain are the features on the plain under glacial Lake Agassiz in North Dakota and southern Manitoba. The dominant characteristic of this plain is its flatness, but rivers have cut very shallow valleys in it. In addition, the sand and gravel deposits that formed as beaches on the edges of glacial Lake Agassiz break the monotony of an otherwise regular surface.

It would be a mistake to think of flat plains as featureless. The detailed features that do exist depend on the kind of tectonic events responsible for the plain and the gradational processes at work on its surface.

ROLLING OR IRREGULAR PLAINS

Flat plains often occur as small parts of much larger plains in which most of the landscape is rolling or irregular. Rolling or irregular plains vary considerably, but have certain definite characteristics in common. These are usually related to the local relief and degree of slope. However, it should be noted that there are wide divergencies in elevation. In addition, the gradational agents are responsible for many differences in both surface material and in the arrangement of landform features.

Running water is the main gradational agent responsible for the detailed character of rolling or irregular plains, just as it is with flat plains. However erosion by running water is more important than deposition by it. In addition, the effect of running water has been complicated, and indeed sometimes surpassed, by other gradational agents. Glaciation is the most notable of these. Large areas of the northern continents owe their present appearance to the effects of both glaciation and running water, as seen in Figs. 8-24 and 8-26. Smaller sections or regular plains have detailed characteristics that are the result of wind and mass wasting.

STUDY 9-2

1 Under the headings (a) flood plains and deltas, and (b) coastal and lake plains, list some of the important flat plains of the world as shown on Fig. 9-1. Where a landform name is not obvious, refer to the location by naming an adjoining body of water, a country, or by latitude and longitude.

2 Using your knowledge of the processes that create landforms, explain (a) how rivers are responsible for the deposition of material sufficient to form a flood plain, and (b) how deposits of sediments form flat surfaces under lakes and in coastal locations.

3 Sketch a map based on Figs. 8-16 and 8-17 and then write a description of the land use in the area depicted, including as much information as possible. Some information on agriculture, vegetation, transportation, and settlement can be found by examining Figs. 8-16 and 8-17. As far as you think reasonable, account for the land use by relating it to the landform characteristics of this type of plain.

4 Many flood plains and deltas are important agricultural areas that support large populations.

a Suggest reasons why the delta and the flood plain of the Mississippi are not as densely populated as some Eurasian ones.
b How have the Dutch made use of the deltas of the Rhine and other rivers?

5 Referring to Fig. 9-1, name and locate the larger rolling or irregular plains of the world. Describe their locations by relating them to their continents and to the oceans they face.

6 From a consideration of some specific examples in Figs. 8-15, 8-29, and 8-31, describe the slope and local relief characteristics of rolling or irregular plains. Mention any notable variations in features between the different rolling or irregular plains used as examples.

7 List plains whose detailed characteristics resulted from:

a erosion by ice and surface water
b deposition by ice and erosion by surface water
c erosion by surface water
d deposition by surface water
e erosion and deposition by wind.

Plateaus and Plains with Hills or Mountains

Plateaus are areas where most of the surface is gently sloping land at relatively high elevation (over 100 m above sea level). They are distinguished from plains by their moderate to high local relief, owing to occasional deep valleys or escarpments. Plateaus are usually dissected by deep valleys or separated into different levels by step-like escarpments. The altitude is often the result of the comparatively recent uplift of a large area *en masse* or of the outpouring of lava over a wide area.

In general a plateau is a youthfully dissected land surface. Given long periods of time this surface will likely be dissected further by surface water, producing a much more irregular, hilly type of landscape. On many plateaus, however, surface water dissection occurs very slowly because it is inhib-

Fig. 9-2. Plains with hills and mountains north-east of Las Vegas, Nevada. How were the hills and ridges in the foreground formed?

ited by one or more of the following factors: low precipitation, resistant surface strata, or permeable surface materials.

STUDY 9-3

1 Referring to Fig. 9-3, describe in words and diagrams the general characteristics of plateaus.

2 Referring to Fig. 9-1, find examples of plateaus in different parts of the world. Why is this landform not nearly as important in total area as most of the other types?

3 With the exception of occasional tectonic structures such as faults or volcanoes, most of the landscape on a plateau is the result of various gradational processes.

a Why have the rivers cut deep valleys or canyons, although their tributaries have not dissected the adjoining uplands? What gradational stage does this represent? Why are plateaus more common in sub-humid areas? What other gradational agents will be involved in forming the characteristics of plateaus?

b The presence of plateaus may be a result of one of the following or a combination of them: the short period of time since uplift or the outpouring of lava occurred; the amount of precipitation; the presence of a

Fig. 9-3. A part of the Colorado Plateau, showing the valley of the Little Colorado River.

protective or permeable layer of surface rock. Describe how each of these can contribute to the presence of a plateau.

c Some of the features of plateaus include canyons, escarpments, mesas, buttes, badlands, and talus slopes. Describe the characteristics of each and the process by which each is formed. Why are plateaus often effective barriers to transportation?

Plains with hills or mountains are areas where the surface consists mainly of gently sloping land at a relatively low elevation, with moderate to high local relief owing to occasional hills or mountains.

Such a landscape originates in one of two ways. The entire surface—at one time hills or mountains—may have been worn down except for occasional more-resistant features. Large areas in the Guianas, in eastern Brazil, and in parts of central and western Africa have been formed in this way. In these areas very resistant crystalline rock has protected some of the former uplands, which now remain as hills surrounded by a fairly level erosional plain. Secondly, plains with hills or mountains may be a result of tectonic activity on a plain. The hills or mountains will then consist of a series of widely spaced volcanoes or, as found in parts of the south western

United States, a series of raised and tilted fault blocks separated by wide valleys.

Hills and Mountains

Mountains are distinguished from hills by their greater local relief—more than 600 m. Mountain slopes are often steep, commonly between 10° and 25°. Generally there will be more rock in the surface material of mountains than in the surface material of any other type of landform. Hills also have moderate to steep slopes and a greater local relief than plains. Despite this, however, there are larger areas of gently sloping land in most hilly regions than in mountain regions.

It is clear from the appearance of mountains that they were created by tectonic activity. Such processes have raised parts of the earth's crust hundreds of thousands of metres above base level. A mountain region may have experienced folding, faulting, and vulcanism, and it may also have been involved in a series of uplifts over a period of tens of millions of years. For example, it is believed that the Rockies have experienced three major uplifts since they came into being at the beginning of the Cenozoic era. It is likely that previous mountains had been worn down before the last uplift only a few million years ago, and that the present mountain peaks represent some of the higher remnants of these previous mountains. Consequently, all major mountain systems are a result of uplift in fairly recent geological time, that is, within the past few million years. However, the detailed features of mountain terrain are a result of gradational activity over a long period of time.

SURFACE FEATURES OF HILLS AND MOUNTAINS

The structure of mountains and hills is very complex. In any one group of hills or mountains, a number of different tectonic pro-

cesses have often occurred in the same place several times over millions of years. Most of the detailed features of mountains are a result of the gradational processes. These detailed features are the aspect of landforms that concern geographers. Consequently, an understanding of how mountains were formed is of less importance to the geographer than the geologist.

There are instances, however, where some indications of the origin of mountains or hills may still remain, at least in the form and pattern of major features. These features may include: a linear pattern of ridges and valleys representing the original synclines and anticlines; a continuous crest line formed by a tilted fault block; and, fairly common, more or less parallel ranges separated by deep valleys or trenches. Volcanic mountains, even when they occur in complex mountain systems, can usually be recognized by their unique structural characteristics.

Hills resemble mountains in many ways but are usually distinguished from them by their low local relief and normally lower elevation. It is difficult and probably of limited value to suggest any specific features that would make it possible to distinguish high hills from low mountains. However, landforms where the local relief is consistently above 600 m would be best classified as mountains.

Hills, like mountains, are a result of tectonic activity. Although hills might seem to be worn-down mountains, this is seldom the case. It is likely that most hills result from such moderate uplifting of the earth's crust that the gradational agents (particularly water) cannot produce large differences in local relief. An example of a region of hills is the ridge and basin province of the Appalachians shown in Fig. 7-17.

STUDY 9-4

1 Terms used to describe or refer to mountains include peak, summit, range, system,

Fig. 9-4. Most of the characteristics of glaciated mountains can be seen in this photograph of a part of the Logan Mountains in the Yukon Territory.

knot, chain, crest line, and cordillera. Using an outside source, explain the meaning of these terms.

2 On an outline map of the world mark in broad strokes the major mountainous areas. Identify the more important ranges and esti-mate the average height above sea level of the general crest line of each. Most of the highest ranges were formed during the Ceno-zoic era. What is the explanation of their great height?

3 In addition to being unsuitable for settle-

ment by large numbers of people, mountains also act as barriers between various parts of the globe. However, some of the mountain passes listed below have been and are today of great importance in the movement of people and goods between various regions. On an outline map mark with arrows the following gaps in mountain ranges and label the regions they connect: the passes in Panama and Nicaragua; the Isthmus of Tehuantepec; the Wyoming Corridor (Oregon Trail); the Dzungarian Gate (southeast of Lake Balkhash in the USSR); the Rhone and Carcassonne Gaps; the Brenner, Simplon, and St. Gotthard Passes of northern Italy; the Khyber and Bolan Passes; and the Crowsnest and Kicking Horse Passes in Canada.

4 It is obvious that mountains are the least habitable of the four major types of landforms. What are some of the ways in which mountainous regions are important to people?

5 Examine the various mountain photographs in this chapter. What evidence is there in any of these of the origin of the mountains?

6 Referring to pp. 253 and 254, prepare an analysis of the mountainous areas shown on Figs. 8-38 and 8-39.

7 Explain why hills are seldom simply worn-down mountains. In what ways would the appearance of worn-down mountains differ from the appearance of hills?

8 Discuss the effects of gradation over a long period of time on the possible relationship between hills, plateaus, and plains that are at a high altitude.

9 An area of hills is shown on Fig. 8-12.

a There are certain difficulties involved in classifying the landforms of this area as hills; in fact, the area could be classified as an irregular plain at a high altitude or as a plateau. Comment on this problem of classification. As geological time progresses, the difficulty will likely be resolved. Explain.

b A part of the area is named the Cape Anguille Mountains. Why is it incorrect to describe these landforms as mountains? There are many instances in other parts of the world where hills are called mountains and vice versa. Why does this happen?

[1]Glenn T. Trewartha, et al., Elements of Geography (5th ed.; New York: McGraw-Hill, 1967), p. 258. Copyright 1967 McGraw-Hill, Inc. Used with permission of McGraw-Hill Book Company.

IO/ MINERALS

Resources are all the natural elements from which people obtain their food and all the other items they use. These natural elements include the basic ingredients of life: air and water; landforms and soil—particularly the elements and minerals contained within the crust; and certain plants and animals. With the exception of minerals, most of these resources have already been discussed in previous chapters.

Resources have a number of things in common. First, they all come only from the thin life layer known as the biosphere. Consequently there is a limit to the total quantity of resources available unless our space or underwater technology increases at an incredibly rapid pace. Thus the earth is still our habitat and resource base, and will continue to be for the foreseeable future.

In the second place, most resources are culturally determined. As one authority[1] stated, 'Resources are not, they become.' A natural element does not become a resource until people perceive a use for it. This presupposes that a society has the technical skills to develop and use the resource, and that it is economically feasible to do so. For example, a primitive tribe will recognize only a relatively small number of resources. A tar sand deposit or a rapidly moving river that could be utilized to generate hydro-electricity

will not be perceived as a resource and in fact may be regarded as a nuisance. Thus, the term resource refers to those parts of the total earth environment useable under present technical, economic, and cultural conditions.

A third important characteristic of natural resources is that they are unevenly distributed over the earth. Not only are economic minerals (see p. 264) such as iron or copper unevenly distributed, but other resources such as fertile soil, commercial forests, and fresh water also vary in quantity and quality from place to place. Two important results of this are the development of major trading patterns among nations, and the existence of rich and poor countries. The uneven distribution of certain resources also plays a major role in international affairs. For example, the keen interest shown by many nations in the economic and political affairs of the countries of Southwest Asia such as Saudi Arabia and Iran is partly related to the vast petroleum resource that this area possesses.

CLASSES OF RESOURCES

There are several ways of classifying resources. First, they can be divided into organic and inorganic classes. The organic class, which is derived from plant and animal

Fig. 10-1.

RESOURCES

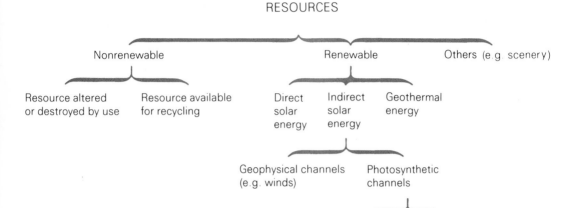

Types of Resources. Based on Peter Haggett, *Geography—A Modern Synthesis* (New York: Harper & Row, Publishers, 1972).

life, includes vegetation, animals, and fish. The inorganic class comprises various earth materials in gaseous, liquid, or solid forms. Air, water, and certain minerals are the most important inorganic materials. Soil falls into both the organic and inorganic classes because it is composed of a mixture of weathered rock fragments and decaying plant and animal materials.

Resources can also be classified according to their availability. Air, water, soil, and natural vegetation are referred to as *renewable* or *flow* resources because they will last indefinitely if managed properly. Unfortunately there are many instances where the quality of a renewable resource has become impaired. This happens when people ignore or do not understand the natural functioning of the biosphere. Two examples are the polluting of air or water, and the depletion of certain fish and animal stocks. The remaining resources, mainly economic minerals, are classified as

nonrenewable or *fund* resources; once they are used up there can be no second harvest. This, however, should be qualified by the fact that many fund resources can be reclaimed and reused over and over again. Reusable resources such as copper, gold, or iron are known as *sustained-fund* resources.

1 Would it be true to say that there are more resources today than at any other time in history? Explain.

2 Using examples, explain how technical, economic, and cultural conditions determine what is a resource.

3 Referring to Fig. 10-1, explain the differences between the various subclasses under renewable and nonrenewable by noting examples for each resource category and how each is used.

ECONOMIC MINERALS

Minerals were defined on page 164 as any naturally occurring inorganic solid with a definite chemical composition and characteristic crystal structure. While this is an acceptable geologic definition it does not satisfactorily cover all those substances occurring in the earth with sufficient value to warrant being mined or quarried. It would be difficult to overstate the importance of mineral resources to the development of civilization. Most of our notable technological achievements have been based on the use of minerals. The terms Copper Age, Bronze Age, and Iron Age, to name periods in the past, indicate both the importance of these substances as well as the length of time they have been used. However, while man has known how to mine, smelt, and make articles from many minerals for over three thousand years, most of the major advances in mineral discoveries—and the development of new technologies based on them—have occurred in the past hundred years. For example, since 1900 mineral consumption has grown by a factor of 12.7, population by 2.4, and per capita consumption by 5.3. When we consider the importance of energy and other raw minerals to our present production technology, it is reasonable to describe minerals as the foundation on which modern industrialization has been built. The resulting high standard of living in industrial countries, particularly those of western Europe and North America, has to a large extent been dependent on the exploitation of minerals.

'Resources are not, they become.' While this statement applies to the concept of resources in general, it is particularly appropriate to mineral resources. Most of the mineral deposits that are presently being used or will be consumed in the future have been in the ground for a much longer time than human beings have been on the earth. The utilization of economic minerals has expanded slowly. First the minerals were recognized; later, uses were perceived for them; and finally, techniques were developed for extracting them from the earth, removing impurities, and manufacturing them into useful articles. For example, in 1500 B.C. about 150 minerals were known, whereas today over 2000 have been identified (although only a small proportion of these have any economic value). All our present economic minerals such as petroleum, nickel, or uranium have been around for millions of years, but they only came to be considered resources in response to human needs and technology. By the same token, there is every likelihood that some mineral known today— but with little or no present economic value —may become an important resource with the development of new technologies.

Our concept of what constitutes an economic mineral deposit has been gradually evolving. Three thousand years ago the first copper mines extracted almost pure or native copper by digging it out of surface rocks on the earth. When these surface deposits ran out the mining stopped. Slowly techniques were developed for taking native or high-grade copper from shallow underground mines. By the 18th and 19th centuries the combination of increased demand and new mining and refining technology led to deeper mines and the extraction of lower grade ores. By 1900 copper smelters could handle ore with about 10 per cent or more copper. Today 0.5 per cent copper is sufficient under most circumstances to justify a mine. Consequently the amount of copper now taken out of the ground is many times greater than at any time in the past. Other minerals such as iron, lead, tin, and gold, which have been mined for thousands of years, have gone through similar stages of development from high- to low-grade ore. As a result of this growth pattern, some economists have suggested that mining will move to lower and lower grade ores as demand increases. It is

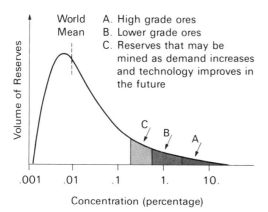

World Mean

A. High grade ores
B. Lower grade ores
C. Reserves that may be mined as demand increases and technology improves in the future

Concentration (percentage)

Fig. 10-2. The theoretical distribution of elements such as iron, aluminum, or copper. Copper is used in the example above. Explain the quantity-quality relationship that this diagram illustrates. (after P. Haggett Geography — A Modern Synthesis).

NONMETALS
I. Water
II. Construction materials—dimension stone, crushed stone, aggregates, gypsum, lime, and cement materials
III. Ceramic materials—clay, feldspar, talc and pyrophyllite
IV. Metallurgical, chemical, and refractory materials—foundry sands, limestone, dolomite, magnesite, phosphorite, fluorspar, sulfur, salts and brine, fire clays, quartz and quartzite
V. Industrial and manufacturing minerals—asbestos, mica, talc, barite, diatomite, graphite, zeolites, bentonite, silica sand, abrasives
VI. Fertilizer materials—sulphur, potash, phosphate, nitrate, agricultural limestone
VII. Gemstones

THE FUELS
I. Coal (including anthracite, bituminous, lignite and peat), petroleum, natural gas, and uranium

assumed that such ores will be increasingly abundant. This is in accordance with a principle called the A/G ratio (arithmetic-geometric). That is, as the grade of ore of certain minerals decreases arithmetically, the abundance of these minerals in the earth increases geometrically (Fig. 10-2). While this principle applies to minerals such as copper, iron, and aluminum, it is not true for many others, including important economic minerals such as lead, zinc, and mercury.

Characteristics of Minerals

Minerals can be divided into three groups: metals, nonmetals, and fuels. The various subdivisions are shown in Fig. 10-3.

Fig. 10-3.

METALS
I. Precious metals—gold, silver, platinum
II. Nonferrous metals—copper, lead, zinc, tin, and aluminum
III. Iron and the ferroalloy metals—iron, manganese, nickel, chromium and molybdenum
IV. Minor metals—antimony, arsenic, barium, calcium, magnesium, lithium, mercury, radium, titanium, zirconium, and other minor metals.

The term 'ore' refers to rocks from which one or more economic minerals can be mined. The term is used mainly in conjunction with the metallic minerals although a few nonmetallics such as sulphur and fluorite may also be described in this way. In each ore body the concentration of the commercial elements is many times greater than the average concentration of that element in the earth's crust (Fig. 10-4). This average concentration of an element in the rocks of the crust is referred to as the *clarke* of the element. Each commercial element has a different clarke. For any rock to be recognized as an ore deposit the concentration of any element must clearly be higher than its clarke and this is called the *clarke of concentration*. Because the amount of a particular element needed to form an ore body varies from element to element, each will have its own clarke of concentration. For example, iron has a clarke of 5 per cent; that is, by weight the average rock of the crust will contain 5 per cent iron. To constitute an ore body the concentration of iron should be at least 30 per cent, thus the clarke of concentration for this mineral is 6 (6 × 5 = 30). Some clarkes and clarkes of concentration for a number of important metals are shown in Fig. 10-4.

Fig. 10-4.

Element	Clarke	Clarke of concentration
Aluminum	8.13	4 ×
Iron	5.00	6 ×
Magnesium	0.10	350 ×
Chromium	0.02	1 500 ×
Nickel	0.008	175 ×
Copper	0.007	140 ×
Zinc	0.013	300 ×
Gold	0.000 000 2	20 000 ×

Some elements such as copper, gold, or silver may occur in a native or pure state. However most *ore bodies* contain minerals in which the valuable element is combined with other elements. For example, copper is usually found in the ore mineral chalcopyrite ($CuFeS_2$) which contains the elements copper, iron, and sulphur; and much of the iron presently mined is found in the ore mineral hematite (Fe_2O_3) or magnetite ($FeO.Fe_2O_3$). In addition, the ore minerals are grouped with other noneconomic minerals in ore deposits. The separation of ore minerals from the ore body leaves a valueless residue called *gangue minerals*. The extraction of the economically valuable elements (metals or nonmetals) from the ore minerals is then carried out through various metallurgical processes which are illustrated in Fig. 10-5. In a concentrator the ore is crushed or separated from the gangue by one of several mechanical or chemical processes. Some of the concentrates, such as construction materials or many non-metallic minerals (asbestos and gemstones), are then ready for use. Most of the metallic mineral concentrates, however, are still in an unuseable mineral form and must undergo further processing. In this second state, called smelting, the concentrate is subjected to high temperatures that cause the impurities

—such as sulphur, oxygen, and carbonate— to separate from the molten metal. While smelting removes most impurities, some metals must go through a further stage known as refining. Increasingly, however, smelting and refining are being replaced by a single hydrometallurgical process in which the mineral content of the concentrate is recovered by leaching and electrolysis. The final product is a marketable commodity. For example, copper and lead will be at least 99.9 per cent pure.

Most countries report a *reserve* figure for the economic minerals within their national boundaries. A reserve is the quantity of a resource that can be economically mined under current prices and with available technology. Reserves are further classified as *proved, probable,* or *possible* according to the extent to which their quantity has been measured. Mineral deposits that are known but cannot be economically mined or are thought to exist but have not been discovered, are described as *resources*. Such resources can also be classified according to the certainty of their existence and the feasibility of their recovery. The main categories include *known marginal, known submarginal, undiscovered marginal* and *undiscovered submarginal.*

Distribution, Formation, and Exhaustibility

The economic minerals of the earth's crust are unevenly distributed. This occurs because of the random nature of the geologic and biologic processes responsible for forming economic minerals. Most mineral deposits are essentially 'freaks of nature' produced by complex geologic processes that occurred under exceptional circumstances. It is important to be aware of the processes that have produced the principal economic minerals because these processes account for their random distribution on the earth. Four of

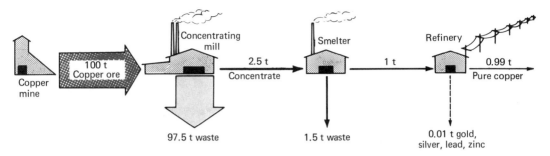

Fig. 10-5. Flow diagram showing the reduction of 100 t of copper ore in the processes of concentrating, smelting, and refining.

these processes are briefly outlined below. (Note: some minerals can be formed by more than one process.)

The major fuel minerals are formed from the remains of plants and animals. For example coal deposits represent the distribution of particular swamp communities during the Carboniferous period (p. 160). Oil and natural gas are formed from the accumulation of plankton remains in the sediment at the bottom of large shallow seas. Most of our present commercial deposits of oil and gas were formed under seas that existed during the Devonian period.

Many of the most important metallic minerals (notably nickel, copper, gold, lead, zinc, and uranium) are associated with igneous intrusions. These economic minerals are found near the surface of the crust, where magma crystallized in the fissures and planes of weakness (cracks). This resulted in higher than average concentrations of certain metallic minerals. Ore bodies produced in this manner are usually found in the earth's shields or in areas of fairly recent tectonic activity.

The concentration of other kinds of economic minerals (ore bodies) may result from the chemical weathering of rocks or regolith. For example, where certain clay minerals in

tropical regions are broken down chemically, they leave concentrations of aluminum hydroxides (bauxite). Other important metallic mineral deposits in tropical and subtropical areas are a result of similar processes. The nickel deposits of New Caledonia are an example.

Mechanical weathering and erosion are also responsible for the high concentration of certain elements sufficient to produce ore bodies. For example, rare metals such as gold or platinum occur as placer deposits. These deposits are produced when the mineralized veins in certain igneous intrusions are weathered. The eroded particles are then carried by rivers and subsequently deposited in stream beds some distance away from their source. Many of the early gold rushes in western North America were focused on such deposits. On a much larger scale, mechanical weathering and erosion have produced concentrations of economic minerals in sedimentary rock. Clays (for bricks or ceramics), phosphate rock, salt, and iron are examples. Much of the world's iron ore comes from such sedimentary deposits (e.g. the Labrador deposits of eastern Canada).

The most important deposits of many of the major minerals are localized. For example, 70 to 75 per cent of the world's non-fuel

Fig. 10-6.

PRODUCTION, CONSUMPTION, RESERVES, AND TRADE OF NINE MAJOR MINERALS 1970[1]

	Percent of world land area	Percent of world's population	Percent of world's mineral production	Percent of world's mineral consumption	Percent of world's mineral reserves	Net trade[2]
Developed Countries [3]	26	22	41	68	35	− 39
Developing Countries [3]	49	50	33	6	38	+ 83
Centrally Planned Countries [3]	25	28	26	26	27	− 2
	100	100	100	100	100	

[1] R. Bosson and B. Varon, *The Mining Industry and the Developing Countries*, published for The World Bank by Oxford University Press, New York, 1977.

[2] The percentage of mineral consumption made up of net mineral imports is referred to as a negative figure (−), and the percentage of mineral production made up of net mineral exports is referred to as a positive figure (+).

[3] A generally accepted division of the world's nations in which the developed group refers to those countries of North America, Europe, Japan, Australia and New Zealand; centrally planned to the USSR, the countries of eastern Europe, China, Vietnam, and North Korea; and developing to most of the remaining countries of the world.

mineral output comes from twelve countries: the USSR (18.7 per cent), U.S.A. (15.9 per cent), Canada (11.5 per cent), Chile (4.6 per cent), Zambia (4.1 per cent), Australia (3.4 per cent), China (3.2 per cent), Zaire (2.4 per cent), Peru (2.3 per cent), South Africa (1.9 per cent), Mexico (1.8 per cent), and Brazil (1.6 per cent). Deposits of individual economic minerals are often associated with specific regions: over 60 per cent of the world's gold comes from an area of less than 2000 km^2 near Johannesburg in South Africa; a large proportion of the world's nickel comes from a number of mines in the Sud-bury area in northern Ontario. The principal deposits of other minerals such as copper, molybdenum, and asbestos are concentrated in a relatively small number of mining districts. It is important to emphasize, however, that some economic minerals—notably iron, bauxite, and many building materials—are widely distributed.

As indicated in Fig. 10-6, the developing countries lag far behind the developed and the centrally planned countries in their consumption and reserves of major non-fuel minerals. It is generally accepted that much of the prosperity of the developed nations is based on economic minerals. These constitute one of the basic resources of a modern industrial state. Consequently, underdevelopment is related to low levels of mineral consumption. This does not mean that minerals are not produced in developing countries, but rather that the bulk of the minerals mined in them is exported to support manufacturing industries in the developed and centrally planned countries.

The geologic events that produce economic mineral deposits occur very slowly. With the exception of certain deposits such as the manganese nodules on the ocean floor, the quantity of minerals available for man's use is fixed. We may be able to recycle much of what has already been mined but the quantities available in the earth are finite. Sooner or later the supply of each of the economic minerals from the earth's crust will run out.

It is important to grasp this relatively simple fact because our rate of mineral exploitation has been increasing year by year. More minerals have been mined in the past four decades than in all preceding history. In fact, as we have just noted, much of the present high standard of living enjoyed by the developed countries is based on an ever-growing consumption of minerals, particularly those used to produce energy. However, a large proportion of the world's population does not possess a standard of living comparable to that in most developed countries. The desire of developing countries to improve their prospects through industrialization will result in an even greater demand for economic minerals in the future. Figure 10-7 shows the reserves, consumption, and estimates of the number of years some particularly important minerals will last.

One further problem should be considered. Mining and related industries have influenced all parts of the biosphere in a variety of ways. On the land, mountains have literally been moved (Fig. 10-8), open-pit coal mines have left permanent scars, and piles of gangue have been left as ugly monuments to past and present mines. In the atmosphere, fumes containing sulphur, lead, ash, and various other substances pollute the air and then are carried into the earth's water systems by precipitation. Many other materials have been discharged directly into lakes and rivers—the deadly mercury is a notable example. Many mining companies are responding to new pollution laws and undoubtedly pollution has decreased somewhat over the past few years. Pollution controls cost the mining industry a great deal of money. Inevitably this affects the price of a product, and the consumer pays for such measures in order to enjoy the benefits of a mineral-based civilization.

STUDY 10-2

1 a For each of the elements listed in Figure 10-4 find the minimum per cent an ore body must contain before the element can be profitably extracted.

b How does Figure 10-4 provide a rough measure of the actual dollar value of the different elements listed?

c The clarke of concentration for each element is only approximate. This value has changed very substantially for certain elements over the past decades and will certainly continue to change. Explain how such factors as the development of new metallurgical techniques, government mining subsidies, demand during periods of war, and increasing demand in general are factors contributing to changes in the clarke of concentration.

2 Just because a particular metal is found in a rock in sufficient concentration to be economically extractable does not mean that mining will occur. What other factors determine whether or not an ore body is commercially exploitable?

3 Explain why the concept of reserves is very

Fig. 10-7

WORLD MINERAL PRODUCTION AND RESERVES

Resource	Known World Reserves 1974	World Production 1974	Estimated Cumulative World Demand 1974-2000	Ratio of known Reserves to Estimated Demand (1974-2000)	Countries with Highest Reserves (% of world) 1974		Major Producers (% of World) 1974	
Bauxite (t)	5.8×10^9	16×10^6	873×10^6	6.6	Australia Guinea Brazil	(21) (17) (17)	Australia Jamaica Surinam	24 20 9
Chromium (t)	523×10^6	2.5×10^6	92×10^6	5.7	Rep. of SA Rhodesia	(74) (22)	USSR Rep. of SA	(31) (23)
Copper (t)	408×10^6	6.5×10^6	320×10^6	1.3	USA Chile	(20) (20)	USA USSR Chile	(19) (16) (12)
Gold (kg)	4.1×10^6	0.08×10^6	3.28×10^6	1.3	Rep. of SA	(61)	Rep. of SA USSR	(68) (17)
Iron Ore (t)	90×10^9	514×10^6	20×10^9	4.5	USSR Canada Brazil	(27) (14) (14)	USSR USA Australia	(24) (11) (10)
Lead (t)	150×10^6	3.1×10^6	123×10^6	1.2	USA Canada USSR	(36) (12) (11)	USA USSR Australia	(14) (18) (11)
Nickel (t)	54×10^6	704×10^3	26×10^6	2.1	New Caledonia Cuba Canada	(22) (16) (15)	Canada USSR New Caledonia	(32) (15) (14)
Tungsten (kg)	1.8×10^9	39×10^6	1.5×10^9	1.2	China	(54)	China USSR USA	(21) (19) (8)
Zinc (t)	236×10^6	5.8×10^6	217×10^6	1.1	Canada USA Australia	(20) (19) (13)	Canada USSR USA	(21) (12) (8)

Natural Gas (m^3)	0.062×10^{15}	1.3×10^{12}	0.05×10^{15}	1.3	USSR Iran	(34) (18)	USA USSR	(46) (20)
Petroleum (t)	118×10^9	3.5×10^9	118×10^9	1.0	Saudi Arabia Kuwait Iran	(20) (14) (12)	USSR USA Saudi Arabia	(17) (16) (15)
Potash (t)	1×10^{10}	24×10^6	1.0×10^9	10.0	Canada E.Germany	(46) (25)	USSR Canada	(25) (23)

SOURCE: *U.N. Statistical Yearbook 1975; Mineral Facts and Problems*, U.S. Dept. of the Interior, 1975.

complex. Explain the relationship between reserves, technology, and economics.

4 Using the information in Fig. 10-6, discuss the relationship between underdevelopment and low levels of mineral consumption.

5 The statistics in Figure 10-7 should provide a basis for discussing world mineral production and reserves. Explain the significance of the figures in column 4. Does a ratio of close to 1 mean that such a mineral will not be available after the year 2000? Explain.

The Metallic Minerals

In the following sections on metallic and non-metallic minerals several essential economic minerals are examined and some conclusions drawn about their distribution, production, reserves, and trade.

IRON AND THE FERROALLOYS

Iron has been described as the 'workhouse metal' of the industrial age. It accounts for over 90 per cent of all metal consumed. Fortunately iron is a common substance with high-grade deposits occurring in all the continents and low-grade deposits in most countries. Consequently the supply of iron based on proven reserves will last for many years. It

is certain that reserve figures will increase in the future as the use of lower-grade deposits becomes economically feasible.

Iron is combined with various metals (called alloys) to make different kinds of steel with specific properties such as strength or resistance to rusting. Manganese is the most important alloy. It is added as a cleansing agent during the smelting process to rid steel of oxygen and sulphur. It also increases the metal's tensile or stretching strength and its resistance to abrasion.

Nickel is another important alloy used in the production of stainless steel and in steels that require strength, stability, and corrosion resistance. In 1975 approximately 40 per cent of the world's nickel was mined in Canada (Sudbury, Ontario and Thompson, Manitoba); most of the remaining 60 per cent comes from the USSR and New Caledonia. Over 90 per cent of the Canadian-produced nickel is exported. The majority of it goes to the United States and smaller quantities to Britain, Norway, and Japan.

ALUMINUM

Aluminum is even more abundant in the earth's crust than iron. As a metal its important qualities include a better weight/

strength ratio than iron, high resistance to corrosion, and good electrical conduction. Most aluminum is derived from bauxite ($Al_2O_3 2H_2O$), a mineral produced by chemical weathering (see page 193) in tropical limestone deposits. The major deposits are relatively young in geologic terms, having been formed in the last 25 million years.

Aluminum refining requires large amounts of electrical power. Consequently aluminum plants tend to be located in the principal industrial countries—notably the United States, the USSR, Canada, and countries of western Europe—in locations where relatively low-cost electricity is available. This situation is changing somewhat as the raw-material producing countries are building refineries in order to reap the benefits of manufacturing their own raw materials. However, it is unlikely that there will be a significant shift in refining capacity as long as the major markets exist elsewhere.

Though bauxite reserves are substantial they are not as great as iron reserves. It is likely that other types of deposits containing aluminum, such as clay, will eventually have to be developed. Most authorities predict that the present technological difficulties in recovering aluminum from clay will soon be overcome. Once this occurs the reserves of aluminum will be almost limitless.

COPPER

Copper was one of the first, if not the first, metals used by man. It is still one of the most important industrial metals principally because of its excellent conducting properties. Until this century most copper deposits were small but high-grade vein deposits (10 per cent or more). These sources have been largely used up and copper of much lower grade (as low as 0.5 per cent) is mined today. It is found in much larger deposits where the copper mineral (usually calcopyrite) is disseminated more or less evenly through large volumes of rock. As a result of this low grade, inexpensive open pit mining practices are often employed (Fig. 10-8).

The bulk of the world's copper reserves are found in six principal producing areas: the western slopes of the Andes in Peru and Chile; northern Rhodesia and the Republic of Zaire; the western United States; the Ural and Kazabestan regions of the USSR; Canada (northern Ontario, Quebec, and British Columbia); and the Lake Superior region of Canada and the U.S.A.

LEAD AND ZINC

These two metals are usually discussed together because their principal ore minerals, galena (PbS) and sphalerite (Zns), are generally found in the same ore body. Lead is used in storage batteries and as an additive in some gasolines. Zinc, a component of brass, is consumed mainly in the production of alloys and in the anti-corrosion treatment of iron and steel. Canada, the USSR, the U.S.A., Australia, and Peru dominate in both the production and reserves of these metals although substantial production of both occurs in Japan, Mexico, Poland, Italy, and Yugoslavia.

OTHER METALS

In addition to the minerals just described, a number of other minerals belong to the metals category. These include tin, silver, the platinoid elements (e.g. platinum, palladium, rhodium, irridium, etc.), mercury, titanium, and gold. Most of these are known as scarce metals because they are found in very low concentration. Their scarcity is shown by their price. In 1974, for example, copper sold for 77 cents a pound, gold $144 an ounce, silver $4.70 an ounce, and platinum $181 an ounce. (All prices in the major international metal markets are quoted in American dollars and Imperial units of measurement.)

Fig. 10-8. Once a hill, the Bingham Canyon open-pit copper mine is now the largest excavation on the earth's surface.

Along with copper, one of the oldest and most interesting metals is gold. The principal producing countries in 1974 were the Republic of South Africa (whose famous Witwatersrand district accounts for over 50 per cent of the world's production) followed by the USSR, Canada, and the United States. Gold is almost universally regarded as an international standard of value and used as a basis for settling international trade balances. Approximately one-half of the world's gold is held by governments and central banks to provide stability for paper currencies. Gold is even held privately as an inflation-proof investment. Gold is also an essential industrial metal used in electronic computers, as a solder in turbine blades in jet aircraft and, of course, in the manufacture of jewellery.

In the late 1960s experts forecast the gradual decline of the gold mining industry because the value of gold had remained fixed at $35 (U.S.) an ounce since 1934. In Canada, for example, production declined by almost 40 per cent during the sixties. Owing to the international monetary crisis of the late sixties and early seventies and the world-wide inflation of the mid-1970s, the price has fluctuated considerably, settling at approximately $175 (U.S.) an ounce by early 1978.

Nonmetallic Minerals

The more important nonmetallic minerals include those used for fertilizers and raw chemicals, and those for building and construction.

FERTILIZER MINERALS

The principal fertilizer minerals include potassium, nitrogen, and phosphorous. They constitute a vital resource in view of the need to increase the world's food production. Fortunately most of the necessary compounds—nitrogen, potash, and phosphorous—are abundant and fairly well distributed.

Potassium (or potash) is an appreciable part of the salt content of sea-water. It has been found in areas once covered by ancient seas. Here the salt formerly in the sea-water was concentrated by evaporation and later covered with other sediments. Important deposits are found in central and northern Germany, the USSR, the Texas-New Mexico area of the United States, and Saskatchewan, Canada.

Nitrogen is quite scarce as a natural salt primarily because it is a relatively inert gas that does not easily unite with other elements. Most nitrogen salts are added to the soil by living organisms. Because they are soluble in water it is only in desert areas that exploitable deposits have accumulated. The main nitrogen deposits (nitrates) are found in the Atacama desert of Chile. Most nitrogen compounds, however, are produced by the electrolytic combination of nitrogen from the atmosphere and hydrogen. In this process the United States is the leading world producer.

The main source of phosphorous is the mineral apatite, which is found in both igneous and sedimentary rocks. In 1974 most of the world's production (U.S.A. 40 per cent, USSR 25 per cent, and Morocco 14 per cent) came from marine sedimentary desposits. The reserves of this element so essential to agriculture are very large.

BUILDING MATERIALS

The largest volume of all mined materials is used as building materials. This group is second only to the mineral fuels in total economic value of all minerals produced. Building materials can be organized into two categories. The first group includes materials that we use more or less as they come from the earth, such as various building stones, sand, and gravel. Second are those that must be processed before they can be molded and set in new forms—clay for bricks, raw materials for cement, and plaster are examples of this type. Most of these materials are readily available and pits or quarries are developed in areas according to demand. The greatest demand, of course, comes from large urban areas. In order to reduce transportation costs, quarries should be located as close to the city as possible. Such quarries can have a destructive effect on the landscape. In many areas stringent regulations have been imposed to reduce the amount of damage done by quarrying.

ENERGY

We have referred to the use of energy many times in this book, emphasizing the role of solar radiation as a part of the natural ecosystem. Here we will be concerned with the use of energy for the purpose of improving the quality of human life.

It would be difficult to overestimate the importance of energy to the development of civilization. Ever since the earliest cave dwellers began to experiment with fire, people have been searching for ways of improving life by the use of energy other than human muscle. The domestication of plants and animals over 10 000 years ago was an important energy-related development that allowed a more efficient use of solar energy through photosynthesis. The next important development occurred almost 800 years later when the water-wheel appeared. Used initially to provide water for irrigation, its power output was gradually improved. By the Middle Ages water-wheels became more efficient and versatile, and could be used in a variety of ways —sawing wood, processing textiles, and grinding grains, to mention a few. The windmill came into common use in Europe by the 15th century and was particularly important in areas that, lacking fast flowing rivers, could not use water-wheels.

By the 17th and 18th centuries limits in the output of energy from traditional sources had been reached while new developments in manufacturing required more powerful and controllable sources of energy. The major break-through occurred in 1776 when James Watt's coal-burning steam engine came on the market. Watt's engine, and subsequent improved versions of it, triggered an industrial and social revolution of immense proportion—a revolution that continues to this day. Indeed, the last two hundred years of history should be interpreted in light of our increasing efficiency in utilizing energy. While the effects of this energy revolution have been more pronounced in the western world —and account for our generally high standard of life—it has influenced all parts of the world. It has also produced problems that have implications for the future existence of the human species on the earth.

Most of the energy used in modern manufacturing originates from solar radiation. In some cases this energy has been stored underground in the form of coal and petroleum. Since the reserves of these can become exhausted, they are referred to as *capital sources* of energy. When the sources of energy cannot be depleted, they are called *income sources*. For example, the influence of solar radiation on the earth's atmospheric envelope causes winds to blow (turning windmills) and rain to become surface run off, which is used to turn water-wheels or produce hydro-electricity. The use of solar heat collectors to produce heat or electricity is another example.

Energy sources can also be classified into *primary* and *secondary* categories. The pri-

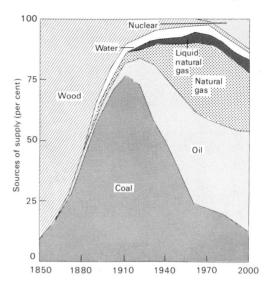

Fig. 10-9. Energy use in North America from 1850 to the present and projected to 2000.

mary group includes those that occur naturally, such as coal, crude petroleum, natural gas, uranium, water, peat, wood, and geothermal sources (energy derived from the heat of the earth's interior). Secondary sources include electricity, manufactured gas, refined petroleum products, and various manufactured solid fuels.

The Mineral Fuels

The developments that initiated the Industrial Revolution in the 18th century were based on the use of coal. Petroleum and natural gas came into widespread use only in this century (Fig. 10-9). All these energy sources are known as fossil fuels.

COAL

Coal is formed from decayed vegetation, principally trees, that existed millions of years ago. Their decay usually occurred in coastal swamps where, lacking oxygen, the plant

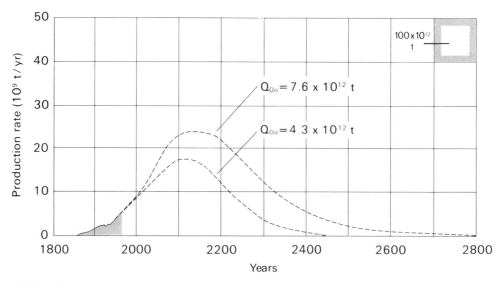

Fig. 10-10. A projection of the complete cycles of the world's coal production. The two values Q∞ repre-sent different estimates of the amount of recoverable crude oil. The shaded area represents the amount consumed up to 1977.

material did not completely decay and later was buried by inorganic sediments. Coal con-sists mainly of carbon, hydrogen, and oxygen, plus certain impurities such as water, ash, and sulphur. Much of our present coal supply was formed in extensive swamps of the Pen-nyslvanian and Mississippian periods (see p. 160). Because of their association with coal, these periods are sometimes combined and called the Carboniferous.

There are several varieties of coal, distin-guished by their content of volatile matter (gas) and their stage of development. *Peat,* the earliest stage, consists of woody plant material found near the ground surface. Over millions of years, pressure and heat from overlying sediments caused the peat to be compressed into a brown coal called *lignite.* Further compression forms *bituminous* coal. Some bituminous coal was affected by both heat and compression and metamorphosed to become anthracite. This black hard coal is

almost pure carbon. The amount of heat a unit mass of coal can produce on burning increases as the stage of development pro-ceeds from peat to anthracite.

After the Industrial Revolution, coal rapidly became an important source of energy and a crucial raw material for manufacturing. Only since the Second World War has petroleum emerged as serious competition. Coal, how-ever, still retains several advantages at pres-ent levels of technology. There is no feasible substitute for it in the smelting of iron ore. Coal is cheaper per tonne in most parts of the world than an equivalent amount of petro-leum (even though a tonne of coal yields less engery). Coal is a flexible source of energy in that it can be used to provide both heat and motive power, the latter either directly in steam-engines or indirectly as thermal elec-tricity. In fact, the generation of electric power is the largest single use of coal. The principal disadvantages of coal—particularly in comparison with petroleum—are (1) lower energy content per unit mass, (2) difficulties in transportation, and (3) the smoke given off during combustion, which includes various offensive impurities (e.g. sulphur) that are major contributors to air pollution.

Although the production of coal in the 1950s and 1960s declined as large supplies of petroleum and natural gas became available, production in the 1970s has been rising. The current energy crisis, combined with the fact that coal is competitive with oil and gas except in home heating and transportation, have caused this. Many experts now feel that coal will help fill the energy void until nuclear and other relatively new forms of energy can be developed. Thus, more coal will be used in the production of electricity. Furthermore, coal can also be used as a source of synthetic oil and gas.

PETROLEUM AND NATURAL GAS

Petroleum and natural gas originated from the incomplete decay of the remains of small marine organisms that accumulated on the sea-floor. These organisms were buried millions of years ago in sedimentary rock—known as *reservoir* rocks—such as sandstone and limestone. They slowly changed into hydrocarbons (petroleum and natural gas) and filled the pore spaces in the rock. Much of the oil we are mining comes from rock of the Devonian age. Normally petroleum deposits are concentrated in localized deposits, called pools. Many of these deposits have been trapped there by some form of tectonic activity. Some of the different kinds of traps are shown in Figure 10-12. Natural gas may be found alone or in association with petroleum deposits.

While small quantities of crude (unrefined) petroleum are used as a fuel or raw material, most petroleum is broken down by refining into various specialized products. Fuels include motor gasoline, kerosene, distillate

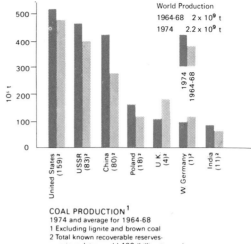

COAL PRODUCTION[1]
1974 and average for 1964-68
1 Excluding lignite and brown coal
2 Total known recoverable reserves-various dates-world 430 (billion tonnes)

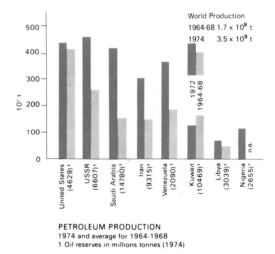

PETROLEUM PRODUCTION
1974 and average for 1964-1968
1 Oil reserves in millions tonnes (1974)

Fig. 10-11. Principal coal and petroleum producing countries.

fuel oil (e.g. diesel oil and light heating oil), and residual fuel oil (e.g. furnace and bunker oil). Non-fuel products of petroleum include various lubricants, asphalt, wax, coke, naphtha, and various petrochemicals. Natural gas is used primarily as a fuel for industrial and domestic uses, in the generation of thermal electricity, and in the production of fertilizer.

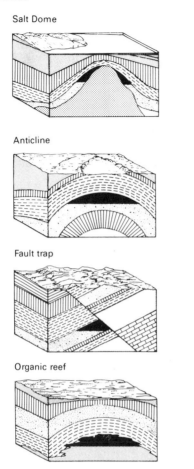

Salt Dome

Anticline

Fault trap

Organic reef

Fig. 10-12. Some types of oil traps. Salt domes and anticlines account for approximately 80 per cent of the total world oil production.

Petroleum has several advantages over coal. First, it is more easily transported. Second, it has a higher energy content per unit mass. Finally, it is a comparatively clean-burning fuel and its combustion is readily controllable. Owing to these advantages, petroleum has taken over many of the tasks for which coal is not well suited. Transportation is the most striking example of an industry now dominated by petroleum.

The world's production and reserves of petroleum are concentrated in a relatively small number of countries. For example, in 1976 the thirteen members of the Oil Producing Exporting Countries (OPEC) had 80 per cent of the reserves and 60 per cent of the production of the non-Communist world. Since many of the industrialized countries such as the United States, Canada, France, West Germany, and Japan are dependent on exports from these countries, 'oil politics' have become important in international affairs.

Natural gas is at present an important source of energy in the United States (47 per cent of world production in 1975), the USSR (21 per cent), Netherlands (7 per cent), and Canada (6 per cent). Its importance is increasing in western Europe with the discovery and development of large reserves under the North Sea. Natural gas is a convenient fuel that can be burned in simple appliances; its combustion leaves no residue and gives rise to little atmospheric pollution. The main disadvantage of natural gas is the high cost of pipelines and supertankers used in its transportation. Gas pipelines must be about four times the size of oil pipelines to carry an equivalent amount of energy. Similarly, tankers that carry liquid natural gas (LNG) must also be very large to carry an economical amount of natural gas. As a result, transport costs have a great influence on natural gas use. Gas fields remote from the market have been of less economic value than those more conveniently located. However, with petroleum prices rising, plans for longer pipelines and the construction of LNG tankers have become more common in the 1970s. The United States began discussion with Canada in mid-1977 for a pipeline across Canada from the Prudhoe Bay gas deposits in Alaska. Similarly, the USSR has announced plans for the construction of a 2000 km pipeline from newly discovered gas fields in western Siberia to the Moscow and Leningrad markets.

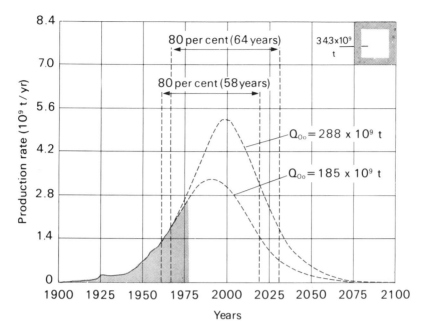

Fig. 10-13. A projection of the complete cycles of the world's crude oil production. See Fig. 10-10 for meaning of the two values of Qoo.

NUCLEAR ENERGY

Since Enrico Fermi developed the first atomic chain reaction in 1942, nuclear reactors have proliferated around the world. The majority of them are used for research purposes. The development of electricity from nuclear reactors began slowly in the 1950s and has accelerated rapidly in the 1970s. Despite this growth, however, the amount of electrical production from nuclear reactors is still fairly small (Fig. 10-14).

Almost all nuclear power operates on the basis of *fission*, the splitting of the nuclei of heavy elements such as uranium. Uranium 235 is the only atomic species that is readily capable of fissioning. Unfortunately uranium 235 is a relatively rare isotope (0.71 per cent of natural uranium). Its total reserves are cur-

Fig. 10-14. Nuclear electric energy production, 1965 and 1974.

Country	1965 10^6 MJ	1974 10^6 MJ	Per cent of total electrical production in 1974
United States	13 165	405 706	5.7
United Kingdom	58 816	121 021	12.3
Japan	90	70 916	4.3
France	3 229	50 155	7.7
Canada	432	49 910	5.0
West Germany	421	43 690	3.9
World	88 920	871 560	3.9

United Nations, *Statistical Yearbook*, 1975.

rently estimated at about 2 000 000 t. Even though reserves will likely increase as new deposits are explored, owing to the rising price of uranium, a shortage of low-cost ores is predicted by the end of the century. This prospect is responsible for the pressure to develop breeder reactors to convert non-fissionable uranium 238 into fissionable isotopes. Uranium 238 (99.3 per cent of natural uranium), is much more abundant than 235. This conversion is already technologically possible. It is estimated that the energy potential obtainable from breeder reactors using low-grade ores of uranium and thorium is thousands of times greater than the energy potential of the world's initial supply of fossil fuels.

Enormous amounts of energy can also be released from the controlled *fusion* of certain isotopes of hydrogen, notably deuterium and tritium; this has already been achieved in the hydrogen bomb. However, before controlled fusion can be used to produce controlled energy, a device must be discovered to contain the fusion reaction. The potential energy available from the fusion process can be estimated even though we do not yet know if, or when, electricity will in fact be produced from fusion reactors. Such estimates indicate a potential millions of times that of the world's initial supply of fossil fuels.

One major problem of nuclear energy is the disposal of radioactive wastes. All radioactive materials are biologically harmful. Many of the most common (e.g. strontium-90, cesium-137) have half-lives of thirty years or less and must be isolated for periods ranging between 600 and 1000 years. The heavier elements must be stored for more than a million years (plutonium-239, for example, has a half-life of 24 600 years). Unsuitable disposal practices, particularly for low-level wastes, are being used in some countries. These include the discharge of low-level wastes into the ocean or their burial a few metres below the surface.

TAR SANDS AND OIL SHALE

Tar sands or oil sands are deposits containing thick tar-like material called bitumen. Bitumen averages about 12 per cent of the total mass of the sand and is made up of oil compounds (50-60 per cent), resins (30-35 per cent), and asphatenes (15-25 per cent). Some of these deposits are found on or close to the earth's surface. Other deposits are covered with varying amounts of regolith, known as overburden.

Tar sands are found in a number of areas in the world. Estimates of petroleum reserves in these deposits vary between 7×10^{10} t and 15×10^{10} t. The Athabaska tar sands in northern Alberta, the largest and one of the most accessible tar sands, is the only deposit in the world being mined on a large scale. The difficulty with the development of this energy resource is the high cost of bringing a plant into production.

Oil shale also exists in large deposits and is found in many parts of the world. The largest occur in the United States where the borders of Utah, Colorado, and Wyoming meet. Oil shale deposits differ from tar sands in that their hydrocarbon content is in solid rather than liquid form. They also differ in chemical content, particularly in the presence of nitrogen (which is expensive to remove in the refining process).

Experiments with the large oil shale deposits in the United States have been going on for many years. Even though the technology exists for processing these shales, a large-scale plant has not yet been built (probably because of the high capital cost of building such a plant).

Electrical Energy

Almost 70 per cent of the world's electricity is generated by burning fuels. Coal and, to a lesser extent, natural gas, petroleum, and uranium are used to drive the steam turbines

Fig. 10-15. The Sir Adam Beck-Niagara Generation Stations Nos. 1 and 2 are located 13 km downstream from the Falls. Construction of plant No. 1 began in 1917 and plant No. 2 in 1950 beside the original Beck plant. Two 9 km underground tunnels, passing directly under the city of Niagara Falls, were built to divert water to the forebay. A reservoir covering almost 290 ha stores water to be released at peak periods to produce electricity. The two plants have a combined capacity of 1 815 000 kilowatts from twenty-six units. A similar system with approximately equal capacity is found immediately across the river in the United States.

that produce thermal electricity. The use of atomic energy to produce electricity is increasing, but as Fig. 10-14 shows it is still of minor importance. Most of the remainder of the world's electricity is hydro-electric, that is, it is generated using water-power. Tidal sources and geothermal sources (which utilize heat from the earth's interior) are of minor importance.

Electricity has the great advantage over coal, oil, and gas of being easily transmitted over long distances by power lines. It requires no storage facilities at an industrial plant, and is convenient to use. Consequently, electricity has largely superseded steam as the source of motive power in most factories. In many areas electricity is also competitive in cost as a simple source of safe, clean heat. The most versatile of all energy commodities, it is used in every country and almost every type of human activity is dependent on it.

Though almost 25 per cent of the world's electricity is produced in hydro-generating plants, it is estimated that this represents only 7.5 per cent of the world's potential hydro capacity. There are several reasons why hydro capacity has remained relatively undeveloped. Much of the hydro potential is located in underdeveloped regions of the world where the high capital costs of constructing hydro generating plants has been an inhibiting factor. In addition, many of the

Fig 10-16.
Production of Electrical Energy, 1974[1]

	Electric energy production (10^9 MJ)	Percentage of electrical production from hydro plants	Approximate production per capita (MJ)
United States	7082	16(− 2)[2]	32 893
USSR	3513	14(− 5)	13 666
Japan	1659	18(− 22)	14 767
West Germany	1122	6(− 4)	18 068
Canada	1004	75(+ 1)	43 470
United Kingdom	984	2(− 1)	17 539
France	694	32(− 13)	12 229
Italy	530	26(− 34)	9 407
Norway	276	100(0)	68 976
India	272	36(− 4)	439
Sweden	270	76(− 4)	32 983
Brazil	254	95(+ 11)	2 304
Australia	251	19(− 6)	18 194
Mexico	147	41(− 3)	2 354
Switzerland	136	77(− 19)	20 866
Austria	122	66(− 2)	16 265
New Zealand	66	77(− 5)	20 646
Iran	45	44(+ 33)	1 321
Ghana	13	99(+ 79)	1 300
Indonesia	12	54(+ 14)	86
Nigeria	10	98(0)	158
Saudi Arabia	5	0(0)	828
World	22 483	25 (approx)	5 616 (approx)

[1]United Nations, *Statistical Yearbook*, 1975.
[2]Figure in parenthesis indicates change since 1965, i.e., in 1965 U.S. was 18%.

undeveloped sites are located in remote regions some distance from areas of demand. The development of hydro facilities in such remote areas is not yet economically feasible because electricity cannot be transmitted more than approximately 1000 km without an excessive loss of power during transmission. Furthermore the development of many sites may be strongly resisted because such development damages or destroys large areas of valley land by flooding. Some of this land is now used for agriculture. Other potential hydro sites would damage some of the most beautiful natural scenery in the mountainous regions of the world. Therefore it is most unlikely that hydro power will be expanded to replace the rapidly depleting fossil fuels.

Energy Consumption

Nowhere has the growth of energy consumption been greater than in the United States and Canada during the 20th century. (Differences in consumption and production of all major regions of the world are given statistically in Fig. 10-17.) The phenomenal growth in the demand for energy of 3.2 per cent annually during this century has been supplied by the exploitation of fossil fuels. Between 1965 and 1970 the annual increase in energy demand was an even more rapid 5.1 per cent. Since the annual rate of population growth averaged only 1.5 per cent in 1900-1970, most of the increased energy consumption reflects rising levels of individual consumption. In the world's greatest consumer society, the relationship between energy supply and demand is less a matter of *how many* people live here than of *how* people live here.

STUDY 10-5

This study provides a basis for discussing several aspects of energy production: its growth over the past twenty years, the situation today, and problems to be faced in meeting future energy needs.

1 Using Fig. 10-17, find the percentage and absolute increase in (a) total energy and (b) total energy consumption and production between 1953 and 1974 for the world's major regions. List the regions in order of their increase in consumption and production. Comment on the significance of your findings. (Various kinds of graphs or simple tables could be used in making this comparison.)

2 Find the per capita energy consumption of each of the regions of Fig. 10-17 for 1953 and 1974. (Divide the total energy consumption by the population.) Which regions showed the greatest increase in this time period? Comment on the change in energy consumption in relation to their population growth and level of economic development.

3 What are the major energy importing and exporting regions of the world? Which of the major energy sources constitute the bulk of the trade? Have any important changes in these aspects of trade in mineral fuels occurred between 1953 and 1974?

4 Is it possible or desirable for all regions to achieve the energy consumption level of North America? In terms of possibility, examine existing and future energy sources, development of technology, the population of the region, and the growth levels in energy consumption since 1953 (see Questions 1 and 2). In terms of desirability, comment on such factors as air and water pollution, optimum levels of energy consumption in terms of quality of life (i.e. do North Americans use more energy than they really need?), and the need in many low energy-consuming regions to find jobs for their population. It is assumed that energy-intensive forms of production (factory manufacturing) will not produce as many jobs as low-energy, labour-intensive forms of production (subsistence agriculture).

5 Using the total world consumption figures

Fig. 10-17

Region	Population 10^6 1953[4]	Population 10^6 1974	Year	PRODUCTION: Coal and lignite	PRODUCTION: Crude Petroleum	PRODUCTION: Natural Gas	PRODUCTION: Hydro and Nuclear	PRODUCTION: Total	CONSUMPTION: Solid Fuels	CONSUMPTION: Liquid Fuels	CONSUMPTION: Natural and Imported Gas	CONSUMPTION: Hydro and Nuclear	CONSUMPTION: Total
Africa	229	391	1953	33	4	0	0	38	33	16	0	0	49
			1974	71	395	13	4	483	71	60	5	4	140
N. America	174	235	1953	454	467	321	22	1264	441	499	333	22	1295
			1974	562	868	879	79	2388	521	1182	873	79	2655
Central and South America	174	212	1953	8	154	7	3	172	10	58	8	3	79
			1974	13	378	56	15	463	17	231	57	16	320
Asian Middle East	48	86	1953	6	155	6	0	161	6	8	0	0	14
			1974	8	1609	49	1	1668	8	74	37	1	120
[2]Rest of Asia	760	1428	1953	89	21	2	6	119	91	29	2	6	128
			1974	126	134	33	18	300	192	435	31	18	677
Europe (except Eastern Europe)	292	341	1953	517	9	4	17	547	523	95	4	17	638
			1974	316	30	221	57	625	376	877	229	58	1539
Oceania	13	21	1953	22	0	0	1	23	22	9	0	1	32
			1974	64	30	7	4	109	39	45	7	4	94
[3]Centrally Planned Economies	910	1170	1953	539	84	13	3	638	531	84	13	3	631
			1974	1332	800	418	26	2576	1291	667	426	26	2409
World	2600	3890	1953	1669 (56%)	895 (30%)	348 (12%)	52 (1.8%)	2963 (100%)	1657	797	361	52	2867
			1974	2497 (29%)	4245 (49%)	1675 (19%)	205 (2.4%)	8621 (100%)	2515	3570	1664	205	7953

Production and Consumption of Energy by Major World Regions Regions 1953 and 1974[1] (10^6 t of coal equivalent).

[1] United Nations, *Statistical Yearbook*, 1975, New York, 1976.
[2] Excluding centrally planned economies.
[3] Includes countries of eastern Europe, the People's Republic of China, Mongolia, Democratic People's Republic of Korea, Democratic Republic of Vietnam, and USSR.
[4] Population figures for 1953 are estimates.

for each of the sources of energy in Fig. 10-17, calculate their percentage of the total world energy consumed in 1953 and 1974. Comment on the changes that have occurred. Why have these proportions changed over this period? Is this a trend that will likely continue over the next several decades?

6 'It is seen that the epoch of fossil fuels can only be a transitory and ephemeral event—an event, nonetheless, which has exercised the most drastic influence experienced by the human species during its entire biological history.'[1]

In what ways has the epoch of fossil fuels '. . .exercised the most drastic influence. . .'? What is the significance of the fact that the epoch of fossil fuels '. . .can only be a transitory and ephemeral event. . .'?

7 With the exception of nuclear fusion and oil sands and shales, there has been little mention in this chapter of new sources of energy that might be used when fossil fuels are depleted. Since this subject has been covered in a variety of sources (some are noted in the bibliography) there is no shortage of information on new developments in energy production. Using such outside sources, examine one or more of the following and note its present stage of development and prospects for the future.

a geothermal
b solar
c bioconversion (organic matter)
d winds
e waves.

[1] M. King Hubbert, 'Energy Resources' in *Resources and Man*, W.H. Freeman and Co., 1969, p. 205.

Appendix
MAPS AND MAPWORK

The making of maps is as old as civilization. In fact even before people could write they made crude drawings on the ground to show locations or directions important to them. As our knowledge of the earth grew, and the need to communicate this developed, the kind of maps people made increased in complexity and variety. However it was not until the first spacecraft photographed the earth's continents that we were able to view large sections of the earth, thus proving the extraordinary cartographic accuracy of our maps.

Cartography involves the preparation of plane (flat) maps, globe maps, models, and diagrams, which are used to represent the earth or a portion of it. The geographer is very concerned with these devices, particularly maps, for they are among the important tools he uses to describe and interpret the earth's variable character. However, in order to appreciate and use maps, it is essential that the geographer understand at least the fundamentals of cartography. This involves the three basic attributes of a map: (1) the scale or area-distance relationship between the map (or globe) and that part of the earth it represents; (2) the type of projection used to present the spherical form of the earth on the two-dimensional surface of a map; and (3) the symbols or cartographic notations used to represent different features of the earth on a map.

Scale

Maps and globes represent the earth or a portion of it in a reduced form. A dimension on a map or globe and its relation to the same dimension on the earth is expressed as a ratio known as the *scale*. For example, a globe with a diameter of 60 cm representing the earth with its diameter of about 12 880 km will have a scale ratio of 60 to 1 288 000 000 (12 880 km × 1000 m × 100 cm) or, more simply, 1 to 21 466 666 or 1:21 466 666 (1 288 000 000 ÷ 60); it can also be written 1/21 466 666. This means that one linear unit on the globe represents 21 466 666 linear units on the earth. This ratio would usually be rounded to a scale of 1:21 500 000. Thus the globe is 1/21 500 000 the size of the earth. The units of a scale can be of any kind— millimetres, centimetres, metres—provided the same unit is used on each side of the ratio. When the same unit is used on each side of the ratio, as 1 cm to 1 000 000 cm, the ratio is called a *natural scale*. Thus one centimetre represents 21 500 000 cm, or one pencil width represents 21 500 000 pencil widths. This ratio or *representative fraction* (R.F.) is a useful shorthand method of stating scale. However, in order to relate measurements made on a globe or map to measurements of the real world, it is usually necessary to convert the ratio into a *scale statement*. The globe with an R.F. of 1:21 500 000 would convert to 1 cm to 215 km (21 500 000 ÷ 100 000, the number of centimetres in a kilometre). One centimetre on the globe therefore represents 215 km on the earth. This procedure is simpler to grasp when working with scales for plans of buildings. House plans are often drawn at a scale of 1:10 or 1:12, which can be readily comprehended and visualized as one centimetre to ten or twelve centimetres.

Scale may also be expressed by using a measured line on a map divided into units representing distances on the earth. This *linear* or *graphic scale* is useful because it is possible to measure distances on a globe or map directly, using a pair of dividers or a piece of paper on which the scale is marked off.

Just as distances on a map or globe are in a scale relationship to distances on the earth, so are areas on a map to areas on the earth. If the scale of a map is 1:50 000, then one *square unit* on the map would represent 2 500 000 000 *square units* on the earth (the R.F. is squared). One consequence of this can be noted by comparing two maps, both 10 cm × 10 cm, one with a scale of 1:25 000, the other with a scale of 1:50 000. It can be seen from the simple calculation above that the map with the scale of 1:25 000 will cover only one quarter as much area as the map with the scale of 1:50 000. This relationship between area and scale is shown in Fig. A-1.

Maps are often described as large-scale or small-scale, but these terms can be confusing. *Large scale maps* (1:1 000 000 or larger; i.e. 1:250 000, 1:50 000,

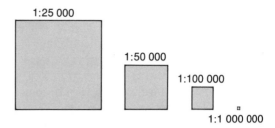

Fig. A-1. An area of approximately 625 m × 625 m on the earth as it would appear on maps of different scales. The size of maps showing the same area of the earth will vary as the square of the ratio of their linear scales.

1:10 000) show a small area of the earth, usually in considerable detail. Topographic maps are drawn at large scales, but objects depicted even on these maps, such a roads and buildings, must be reproduced out of scale in order to be recognizable. Maps on which objects are shown almost at true scale are usually called *plans* and are drawn at scales of 1:10 000 or larger. *Small-scale maps* (1:1 000 000 or smaller, i.e. 1:5 000 000, 1:8 000 000) show relatively large areas of the earth on one map. They are usually found in atlases or books. Many are *thematic* maps which, because of their small scale, are restricted to representing a limited range of information.

STUDY A-1

1 When making measurements on a map, a representative fraction is not particularly useful until it has been changed into a scale statement. If this conversion is necessary, why are representative fractions used? Listed below are a number of R.Fs. Change each of these into an appropriate scale statement. 1:316 000; 1:10 000; 1:63 000 000; 1:50 000; 1:8 000 000; 1:1056.

2 Find the representative fraction for the following scales: 1 cm to 10 km; 1 cm to 200 km; 4 cm to 1 km; 1 mm to 100 cm.

3 Draw a linear scale in kilometres for a scale of 1 mm to 1 km.

4 On a map of unknown scale the corner posts of a straight fence measure 2.7 cm apart. If the actual ground distance between them is 0.6 km, find the representative fraction scale for the map.

5 A map has an R.F. of 1:1000. How many metres on the ground are represented by 4.3 cm on the map?

6 A piece of property 5 km × 5 km is shown on two different maps, one with a scale of 1:50 000 and the other 1:200 000. Find the dimensions and area of this property on both maps (in centimetres and in square centimetres). How many times larger will this property be shown on the larger scale map?

7 Maps A and B have the same scale. A shows twice as much land area as B. How will the two maps differ?

8 Some years ago Canada's National Topographic service changed the map scales of their topographic map series. For example, the map series at the scale 1:63 360 became 1:50 000. Give reasons for the change.

MAP PROJECTIONS

The purpose of map projections is to overcome the problem of representing the earth's spherical surface on a flat piece of paper. The surface of a sphere cannot be transferred to a plane or flat surface without stretching, shrinking, twisting, or tearing, which causes distortion. A map projection is an orderly system of parallels and meridians that controls the nature and extent of distortion and restricts maximum distortion to the least important parts of the map. Only globe maps are free of distortion because they are scale models of the surface of the earth with a similar curved, three-dimensional surface.

There are many different types of projections, each with different properties. No one projection can be perfect, although on large-scale maps distortion may be no greater than the width of a line. The choice of a particular projection depends on the purpose of the map because a particular projection will have certain characteristics shown correctly (e.g. shape) and other characteristics that are distorted (e.g. area). The job of the cartographer, therefore, is to choose the best type of projection for the geometrical requirements of the proposed map.

The easiest way of visualizing the basic principle of projections is to imagine a globe made of glass with the parallels and meridians represented by black lines. Inside the globe a small light bulb at the centre of the earth will project the shadows of the meridians and parallels on to a piece of paper held against the globe. Many different kinds of projections can be obtained in this way depending on the location of the light and the position and shape of the paper against the globe. Some projections are constructed using the principle of projected light. These are known as *perspective projections*. Others are made following the same principle but with some modification, while the largest group are constructed on the basis of mathematical formulas.

There are three main classes of perspective projections: (1) zenithal (or azimuthal), in which the parallels and meridians are projected on to a plane; (2) cylindrical, in which they are projected on to a cylinder; and (3) conical, in which they are projected on to a cone. (Although these are classed as perspective projections, they are usually modified to some extent in order to minimize distortions.) These three are based on the principle of the projection of the meridians and parallels on to a piece of paper. For example, both cylinders or cones can be placed around the transparent globe noted

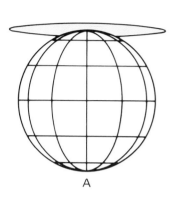
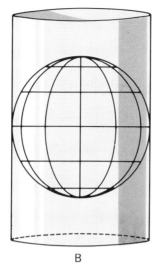
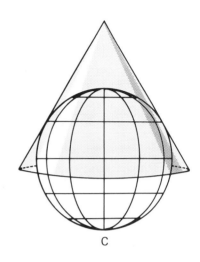

Fig. A-2. Different surfaces on which the earth's grid may be projected. The origin of the projecting lines may be located at the centre of the earth, at a point opposite the point of tangency, or at infinity. A is a projection to a plane; B is a projection to a cylinder; C is projection to a cone.

earlier and then flattened out on a plane surface without any distortion. The perspective projections are illustrated in Fig. A-2. The fourth, largest, and most commonly used class of projections is constructed on the basis of mathematical formulas. These are not true projections in the original sense of the term, as used by mathematicians. They are still referred to as projections because they develop projection-like properties and serve the same function as true projections.

While it is useful to understand the construction of these projections, the essential need for geographers is to be able to recognize the various qualities of each projection and to chose the best one for the purpose of the map in hand. For maps of a continental or world scale the simplest way to determine the characteristics of different projections is to examine the shape of areas bounded by parallels and meridians at different latitudes *on the globe*. When these shapes are compared with the same area on the projection the type of distortion is readily apparent.

SYMBOLS

The map and all its parts are a symbolized representation of the earth's surface. The line representing a coastline on the map is a symbol used to represent the boundary between land and water. On the earth it is not a line at all. Similarly, cities are often represented by squares or circles. The information on maps is chosen according to the purpose of the map, and symbols are then used to represent this information. A photograph is simply a record of a portion of the earth as seen through the camera lens. No symbols are used and all visible aspects of the area photographed can be viewed.

Most of the symbols are readily recognizable. There is, however, one particular type of symbol, the *isarithm*, that requires further explanation. An isarithm is a line that joins points having the same numerical value (its name derives from the Greek *isos*, meaning equal, and *arithmos*, meaning number). The individual lines are often named by combining the prefix *iso-* with a name derived from the type of data—for example, *isotherm* (temperature), *isobar* (pressure), *isohyet* (precipitation), and *isohypse* (elevation above sea level). Isohypses are better known as *contours* and are used to show differences in elevation. Thus a *contour line* is one that passes through points having the same elevation above sea level. This type of line and the topographic maps on which it is usually used are discussed in the next section.

The difference between two adjacent isarithms on the same map is known as the *interval* (the rate at which a quantity changes is known as the *gradient*). When preparing a map using isarithms, careful consideration must be given to this interval. Very small intervals may make a map difficult to read and may also give an impression of accuracy that is not warranted by the amount of information used to construct it. On the other hand, too large an interval may not show enough detail to make the map worthwhile.

When constructing isarithmic maps from plotted data as shown in Fig. A-3 it is often necessary to *interpolate*. This simply means that the isarithms will have to be plotted not only joining points that have the values they represent, but also drawn in proper relationship to values that are greater or smaller by less than one interval. The principle involved is shown on Fig. A-3.

Contour Lines

The problem of representing the three dimensions of the earth's terrain on maps is one of special concern to both the cartographer and the geographer. Many different techniques have been used, but contours, first employed in the early 18th century, are the most common and also the most useful.

Figure A-3 illustrates how contour lines are drawn. Note how the four contour lines are interpolated. The contour interval, the difference in elevation between consecutive contour lines, is 25 m. Why would it be unrea-

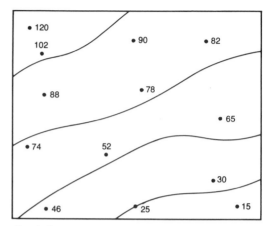

Fig. A-3.

sonable to choose a 10 m interval on the basis of the spot heights shown on Fig. A-3?

When interpreting contours, it is important to remember four basic rules:
1 The lines are continuous; they never cross or branch.
2 If a line closes on itself, then the land slopes upwards from the area outside the contour line to inside the enclosure.
3 Lines that enclose depressions are specially marked to show that the area within the contour lines is lower.
4 Lines approaching a stream valley always bend in an upstream direction.

These characteristics and others can be seen by examining the block diagram and contour map of Fig. A-5.

STUDY A-2

1 Using a sheet of tracing paper, complete the 100 m contour lines on Fig. A-4.
2 Draw contour maps, each 8 cm², to show the following terrain features:
    a a ridge;
    b an escarpment;
    c a river with several tributaries flowing in a wide valley;
    d a uniform, a concave, and a convex slope.

Fig. A-4. Contour exercise No. 1 (after Strahler).

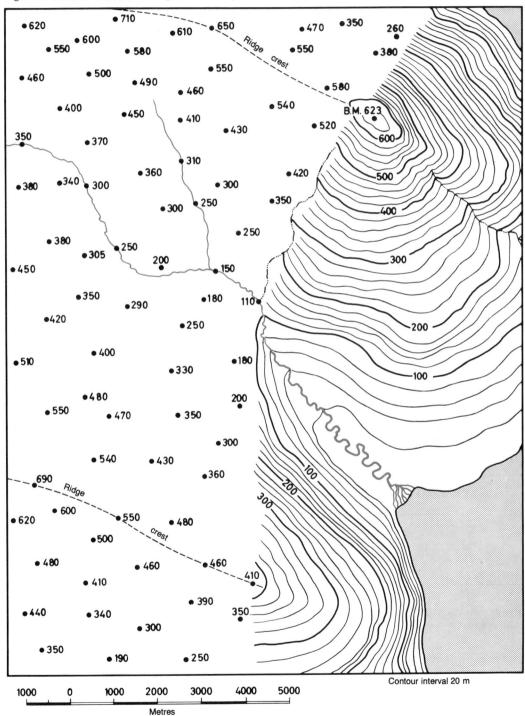

Contour interval 20 m

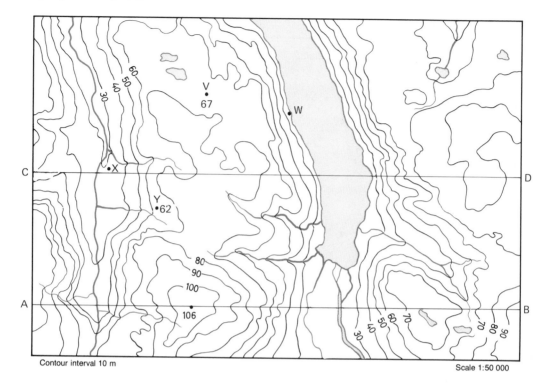

Contour interval 10 m

Scale 1:50 000

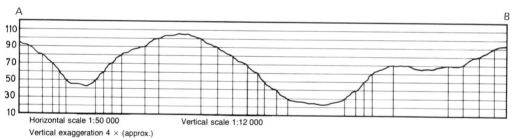

Horizontal scale 1:50 000   Vertical scale 1:12 000

Vertical exaggeration 4 × (approx.)

Fig. A-5.

Profiles

The drawing of profiles is of great value in understanding contours and the representation of landforms. It should, therefore, be a part of any attempt to interpret a difficult contour map. The line along which the profile is made is usually straight, although it can also be drawn along such irregular lines as a road

or a river. In general, if the profile is intended to reveal the character of the terrain, i.e. the degree of slope, amount of sloping land, elevation, etc., it is best to draw the profile across the grain of the landscape, i.e. at right angles to the general direction of the contour lines. The following steps and Fig. A-5 explain the procedure for constructing a profile.

1 Mark the line of the profile (AB) on the map

and transfer this line to a blank page to form the base line of the profile. (The profile can also be constructed on ordinary lined paper.)

2 Construct a vertical scale for the profile. It is almost always necessary to exaggerate this in relation to the horizontal scale. The degree of exaggeration necessary will depend on the scale of the map and the nature of the terrain being portrayed. In Fig. A-5, which allows 2.5 mm for 30 m, the exaggeration will be a little more than 4 times.

3 Place the edge of a piece of paper along the line of the profile. On this piece of paper mark A and B and tick the point at which each contour line touches the edge of the paper. Below each point indicate the value of this contour line. Transfer the piece of paper to the base line of the cross section, putting a mark on the base line opposite each mark on the piece of paper. Then, noting the value of each mark, place a dot above it at the proper height according to the vertical scale; a line joining these dots will form the profile. Note that the line should not be a smooth one. Where two contour lines of the same value are side by side, as in a river valley, the line connecting them should not be straight but should dip down slightly, indicating that the elevation of the river itself will be slightly lower than the elevation of the two adjacent contour lines.

Once you can read contours and make profiles by the above method, it should be possible to make rough profiles quickly and still retain some degree of accuracy. To do this, a vertical and horizontal scale are still necessary. However, the contour lines can be read directly from the map and transferred by eye to the profile. Only important contour lines need be used, i.e. those marking the tops of hills or the bottoms of slopes.

An accurate block diagram can be made by combining two adjacent profiles. The profiles must be parallel (or close to being so) and of the same length. When combining the two, care should be taken to set one above and slightly to the right of the other so that the proper three-dimensional effect can be obtained.

Gradients

It is useful to be able to calculate the degree of slope or gradient from a contour map. Slope is a vital factor affecting such matters as land use, soil erosion, the distribution of settlement, and transportation. In most instances slope is more important than mere elevation.

The gradient of any slope can be expressed as a ratio, or this ratio can be converted to degrees. In the first instance (as shown in Fig. A-6), the vertical distance between A and B is put over the horizontal equivalent, the straight line distance between A and B on the map.

$$\frac{\text{Vertical Distance}}{\text{Horizontal Equivalent}} = \frac{200}{1000} = \frac{1}{5}$$

Thus this slope can be expressed as 1 in 5, meaning that for every 5 m along the horizontal there is a 1 m rise.

To convert this ratio to degrees, first change the fraction to a decimal; then, using tangent tables (an abbreviated version is shown in Fig. A-7), the value of the decimal can be found in degrees. Thus 1 in 5 is 0.20, which is approximately 11°.

Fig. A-6.

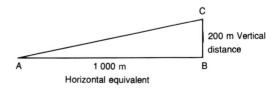

C

200 m Vertical distance

A — 1 000 m — B

Horizontal equivalent

Angle	Tan
1°	0.0175
2°	0.0349
3°	0.0524
4°	0.0699
5°	0.0875
6°	0.1051
7°	0.1228
8°	0.1405
9°	0.1584
10°	0.1763
12°	0.2126
15°	0.2679
20°	0.3640
25°	0.4663
35°	0.7002
45°	1.0000

Fig. A-7.

The previous discussion of profiles and gradients has used the example of contours representing surface configuration because the three dimensions of the earth are readily apparent to us. Other physical and human phenomena exhibit similar characteristics, although they are not always visible. In Chapter 3 pressure gradients are discussed. We can also think in terms of land-value gradients and profiles in a city or a rural area, gradients of population density or temperature and precipitation. Anything that can be represented by isarithms can also be explored by constructing profiles or calculating gradients. The gradient, after all, is only an expression of rapidity of change with distance and the profiles assist us by showing this change in three dimensions. It is frequently helpful when interpreting patterns of isarithms representing any phenomena in geography to think in terms of a three-dimensional surface with gradients of different steepness and high and low areas of different magnitude.

1 Make an accurate profile along CD on Fig. A-5 with the same scales that were used to construct the profile along AB. Combine the two profiles to make a block diagram. On this diagram a number of contour lines can be drawn to represent the terrain between the two profiles. Other information, such as the location of the lake and the rivers, should be shown.

2 Calculate as a ratio and in degrees the gradient of the river that flows parallel to the left-hand margin of Fig. A-5.

3 On Fig. A-5 find which slope is steeper, that between X and Y or that between V and W. Find the gradient of each slope in degrees.

TOPOGRAPHIC MAPS

The word 'topography' is commonly used as a synonym for terrain or surface configuration of the earth. Since terrain is one of the more important characteristics of places, this restricted meaning of topography has wide acceptance. Topography also has a more general meaning, that is, a detailed description of some of the physical and cultural aspects of a small area. It is in this sense that the adjective is used in 'topographic map'. Topographic maps include a wealth of information: the network of road, rail, and streams; the location and sometimes names of important buildings; the areal boundaries of settlements; various categories of vegetation; the location of some primary activities such as mines; the boundaries of administrative divisions; and so on.

Topographic maps are therefore medium- to large-scale representations of a small part of the earth. In order to show a variety of features their scales are always larger than 1:1 000 000 and usually larger than 1:250 000. The most useful topographic maps for many geographic problems are those at the 1:50 000 and 1:25 000 scales.

The Interpretation of Topographic Maps

Like all maps, topographic maps are symbolic representations. They use symbols to show a two-dimensional model of the earth's surface, and in order to read them effectively it is necessary to be able to interpret the symbols. Because topographic maps are drawn at a large scale their symbols more closely resemble the object they represent than those on maps of a smaller scale.

SINGLE-TOPIC MAPS

In order to simplify the analysis of a topographic map it is often useful to isolate one or two phenomena on a topographic map by drawing single-topic maps. In order to draw single-topic maps it is necessary to reduce the scale of the topographic sheet so that the map can be shown on a notebook-size page. This can be done by using some of the lines of the grid of the topographic sheet or by drawing lines if they do not exist. Such a grid, when reduced to a convenient size, can be used to transfer selected information from the original topographic sheets. Sometimes selected items of information may be extracted from topographic maps and presented as a single-topic map for a particular purpose. At other times, when the aim is a more comprehensive analysis of the total character of an area, a series of single-topic maps can be drawn, each summarizing an important characteristic of the area. Normally they would include maps of the following elements: landforms, vegetation and rural land use, transportation, and settlement.

In this way the patterns of such objects and the areal association with other phenomena become more apparent. The observation of spatial associations of things on maps is often one of the initial steps in recognizing that one pattern may partly cause the other. Such associations are difficult to identify by observation of all features shown on the topographic map at the same time and even more difficult when the real world is directly observed.

AIR PHOTOGRAPHS

The development of aerial photography has been perfected only in recent decades. It has been of enormous importance to cartography and geography, as well as to many other sciences. The great value of air photographs is that they provide a visual record of features of the landscape in their true relationship. However, there are also disadvantages. An air photograph is not a map; it provides a view of a part of the earth's surface but without selection of symbols. Thus the interpretation of photographs requires trained judgement.

Air photographs generally fall into two categories. *Vertical* views (see p. 212) are made with the optical axis of the camera perpendicular to the horizontal plane of the ground, or as nearly so as possible. *Oblique* views (see p. 260) are made with the optical axis forming an angle of less than 90° with the ground. On a high oblique photograph the horizon is visible, while on a low oblique it is not.

Interpretation of air photographs is difficult because they show an image of part of the earth's surface taken from an unaccustomed viewpoint (this is particularly true of the vertical photograph). Furthermore, most air photographs are in black and white with all the intermediate tones of grey. Thus many features on the photograph must be identified by their tone, and this tone is largely dependent on the amount of light reflected by the object on the ground. Different types of crops or tree species are examples of phenomena that can be identified by tone (see Fig. 8-14), but this requires considerable skill.

It would take many pages to list all of the uses made of air photographs. Some of the more common ones (geographic and non-geographic) include studies of land utilization, the assessment of timber volume, studies of soils and underlying geology, the planning of road and rail routes, sites for dams and irrigation channels, and so on. One interesting application has been in the field of archaeology. Old field patterns, settlements, and roads that are almost invisible on the ground, can often be seen quite clearly when photographed from the air.

Characteristics of Air Photographs

Most air photographs are taken in a series of parallel strips. Flight lines are organized so that the adjacent strips overlap, usually by 40 per cent. Along each flight line the photographs are taken so that there is also a forward overlap, normally 60 per cent. This is done for a variety of reasons, one of which is to provide two views of the same piece of ground so that it can be viewed stereoscopically.

The scale may vary between two air photographs taken consecutively, and even from one part of the same photograph to another. The first type of variation is usually caused by fluctuations in the path of the aircraft, while the second is caused by *parallax*. In this phenomenon objects that stand above the general elevation of the ground, such as buildings or hills, will be displaced outwards on the photograph, while objects below the general elevation, such as valleys or roadcuts, are displaced inwards. Despite these irregularities, a nominal scale for vertical photographs can be calculated. The scale is a function of the height of the camera above the ground (H) and the focal length of the camera (f) where H and f are in the same units. Thus the scale as a representative fraction can be calculated by the formula R.F. = $\dfrac{f}{H}$. As shown on

Fig. A-7, the scale would be $\dfrac{150 \text{ mm}}{2400 \text{ m}} =$ $\dfrac{150 \text{ mm}}{240\,000 \text{ mm}} = 1:16000$. Using the scale the ground area encompassed by any vertical photograph can be calculated in the following way:

Length (and width) in metres = $\dfrac{\text{denominator of R.F.} \times \text{length of photo (cm)}}{100}$

Stereophotographs

People observe objects simultaneously from both eyes because they have binocular vision. As a result our brain receives a pair of slightly different images—one from each eye. This situation is called retinal disparity. When we view a photograph we perceive depth by means of several clues—relative size of objects, linear perspective, and interposition of objects. However, a single photograph, like a painting, cannot reproduce the visual world as we see it with our binocular vision. When viewed by stereoscopic glasses, however, pairs of pictures taken of the same area from slightly different points of observation are blended, thus permitting us to see the third dimension.

Stereovision is the principle by which an observer can view objects in a photograph in three dimensions (Fig. 8-14). As mentioned earlier, air photographs are taken at intervals that allow a 60 per cent forward overlap. Thus two adjacent pictures will show some of the same ground but from different angles. When they are placed side by side and the right eye is focused on a specific area in one photograph and the left eye on the same area in the other photograph, the change in parallax between the two photographs enables one to see the third dimension. An instrument called a *stereoscope* is commonly used to limit the vision of each eye to its photograph and to provide some magnification. Two rules should be followed when looking at air photographs with a stereoscope: each photograph being viewed should be directly under one eye, with the two areas parallel to the eye base; and the shadows cast by the objects in the photographs should be towards the viewer. It should also be noted that the vertical dimension is considerably exaggerated, producing what is known as relief stretching. This occurs because the average distance between the pupils is approximately 6 cm while stereopairs are exposed from positions in the sky usually several tens of metres apart. However, there is no simple way to measure the amount of exaggeration that occurs unless the height of one object is known. Only then can the height of others be judged in relation to the height of the known object.

Stereophotographs are very useful to anyone interpreting a landscape. Since large parts of the world have been photographed from the air, and stereographic pairs are available for many regions at a relatively low cost, the study of small areas should include the use of such photographs. Stereophotographs have also been invaluable in the construction of contour maps. Almost all contour maps are now produced from photographs. This method is not only simpler and more accurate than the methods previously used, but it also facilitates the mapping of large, relatively inaccessible areas, such as many parts of northern Canada.

Photographic Interpretation

The technique of photographic interpretation is much the same as that used with a topographic map. However, it must be emphasized that photographic interpretation is generally more difficult than the interpretation of topographic maps. Actual features are often difficult to identify from a height of several thousand metres; even distinguishing between a road and railway can present problems initially. With a little practice, however, it should be possible to analyse the area covered by a photograph and discover information about the following: landforms, vegetation, drainage, the area devoted to agriculture (distinguishing between cropland, pastureland, and land in orchards), the pattern of communications, and the distribution and characteristics of settlements.

STUDY A-4

1 a How does the scale of a photograph change as the altitude at which it is taken decreases? Draw a diagram to explain your answer.

b Find the scale of a vertical photograph taken by a camera with a 200 mm focal length at an altitude of 1.5 km. Express the answer both as a repesentative fraction and as a scale statement.

BIBLIOGRAPHY

GENERAL REFERENCES

Ehrlich, A.H., P.R. Ehrlich, and J.P. Holdren. *Ecoscience: Population, Resources, Environment.* San Francisco: W.H. Freeman and Company Publishers, 1977.

Haggett, P. *Geography: A Modern Synthesis.* New York: Harper & Row, Publishers, 1974.

Kendal, H.M., R.M. Glendinning, R.F. Logan, and C.H. MacFadden. *Introduction to Physical Geography* New York: Harcourt Brace Jovanovich, Inc., 1974.

Simmon, I.G. *The Ecology of Natural Resources.* London: Edward Arnold, 1974.

Strahler, A.N., and A.H. Strahler. *The Earth Sciences.* New York: Harper & Row, Publishers, 1971.

––––––. *Environmental Geoscience.* Santa Barbara: Hamilton Publishing Co., 1973.

––––––. *Elements of Physical Geography.* New York: John Wiley & Sons, 1976.

Trewartha, Glenn T., A.H. Robinson, and E.H. Hammond. *Fundamentals of Physical Geography,* 3rd ed. New York: McGraw-Hill Book Company, 1977.

SPECIFIC REFERENCES

Andrews, W.A. (ed.). *A Guide to the Study of Environmental Pollution.* Toronto: Prentice-Hall, Inc., 1972.

Association of American Geographer Resource Papers. Bryson, R.A., and J.E. Kutzback. *Air Pollution.* Resource Paper #2, 1968.

Tuan, Y. *Man and Nature. Resource Paper #10, 1971.*

Hewitt, K., and F.K. Hare. *Man and Environment.* Resource Paper #20, 1973.

Barry, R.G., and R.J. Chorley. *Atmosphere, Weather, and Climate.* Metheun & Co., 1976.

Becht, J.E., and L.D. Belzung. *World Resource Management.* Englewood Cliffs: Prentice-Hall, Inc., 1975.

Bosson, R., and B. Varon. *The Mining Industry and the Developing Countries*. Published for the World Bank by Oxford University Press, Inc., New York 1977.

Bunnett, R.B. *Physical Geography in Diagrams*. London: Longmans, 1975.

Burton, I. *The Environment as Hazard*. New York: Oxford University Press, 1978.

Chorley, R.J. (ed.). *Water, Earth and Man*. London: Metheun, 1969.

Christy, F.T., and A. Scott. *The Common Wealth in Ocean Fisheries, Some Problems of Growth and Economic Allocation*. Published for Resources for the Future Inc., by The John Hopkins Press, Baltimore, Maryland, 1965.

Committee on Resources and Man. *Resources and Man*. San Francisco: W.H. Freeman and Co., 1969.

Dasmann, R.F. *Environmental Conservation*. New York: John Wiley & Sons, Inc., 1972.

Detwyler, T.R., and M.G. Marcus. *Urbanization and Environment*. Scituate, Mass.: Duxbury Press, 1972.

Eyre, S.R. *Vegetation and Soils*. London: Arnold, 1963.

Fairbridge, R.W. *The Encyclopedia of Oceanography*. New York: Van Nostrand Reinhold Company, 1966.

Flawn, P.T. *Mineral Resources*. Chicago: Rand McNally & Company, 1966.

Foster, R.J. *General Geology*. Columbus: Charles E. Merrill Publishing Company, 1973.

Francis, P. *Volcanoes*. New York: Penguin Books Inc., 1976.

Garner, H.F. *The Origin of Landscapes*. New York: Oxford University Press, 1974.

Gass, I.G., P.J. Smith, and R.C. Wilson. *Understanding the Earth*. Cambridge, Mass.: The M.I.T. Press, 1971.

Gilluly, J., A.C. Waters and A.D. Woodford. *Principles of Geology*, 3rd ed. San Francisco: W.H. Freeman & Co., 1968.

Griffiths, J.F. *Applied Climatology*, 2nd ed. Oxford: The Clarendon Press, 1977.

Hallam, A. *Revolution in the Earth Sciences*. Oxford: The Clarendon Press, 1973.

Hardin, G., and J. Baden. *Managing the Commons*. San Francisco: W.H. Freeman and Company Publishers, 1977.

Hare, F.K. and M.K. Thomas. *Climate Canada*. Toronto: John Wiley and Sons, Inc. 1974.

Hewitt, K. *Lifeboat: Man and a Habitable Earth*. Toronto: John Wiley and Sons, Inc., 1976.

Hickling, C.F. *The Farming of Fish*. Elmsford, N.Y.: Pergamon Press, Inc., 1968.

Janes, J.R. *Geology and the New Global Tectonics*. Toronto: Macmillan Co., Inc., 1976.

Lockwood, John G. *World Climatology: An Environmental Approach*. New York: St. Martin's Press, 1974.

McCormick, J.M. and J.V. Thiruvathukal. *Elements of Oceanography*. Philadelphia: W.B. Saunders Company, 1976.

MacNeill, J.W. *Environmental Management*. Queen's Printer, 1971.

Manners, I.R. and M.W. Mikesell (eds). *Perspectives on Environment*. Washington: Association of American Geographers, 1974.

Mather, John R. *Climatology: Fundamentals and Applications*. New York: McGraw-Hill Book Company, 1974.

Morisawa, M. *Streams: Their Dynamics and Morphology*. New York: McGraw-Hill Book Company, 1968.

Odum, E.P. *Fundamentals of Ecology*. Philadelphia: W.R. Saunders Company, 1971.

Ophuls, W. *Ecology and the Politics of Scarcity*. San Francisco: W.H. Freeman and Company Publishers, 1977.

Packard, G.L. *Descriptive Physical Oceanography*. Elmsford, N.Y.: Pergamon Press, Inc., 1975.

Peach, W.N., and J.A. Constantin. *Zimmerman's World Resources and Industries*. London: Harper & Row, Publishers, 1972.

Pirie, R.G. (ed.). *OCeanography: Contemporary Readings in Ocean Sciences*. New York: Oxford University Press, 1977.

Polunin, N. *Introduction to Plant Georgraphy and Some Related Sciences*. New York: McGraw-Hill Book Company, 1960.

Putnam, W.C., and A.B. Bassett. *Geology*. New York: Oxford University Press, 1971.

Robinson, A.H. *Elements of Cartography*. New York: John Wiley & Sons, Inc., 1969.

Ruedisili, L.C. and M.W. Firebaugh (eds). *Perspectives on Energy*. New York: Oxford University Press, 1975.

Russell, E.J. *The World of the Soil*. New York: William Collins Sons & Co., Ltd., 1967.

Scientific American, 'The Biosphere', Vol. 223, No. 3, September 1970.

Shelton, J.S. *Geology Illustrated*. San Francisco: W.H. Freeman and Company, 1976.

Skinner, B.J. *Earth Resources*. Englewood Cliffs, N.J.: Prentice Hall, Inc., 1969.

Smith, Keith. *Principles of Applied Climatology*. New York: McGraw-Hill Book Company, 1975.

Tank, R.W. *Focus on Environmental Geology*. New York: Oxford University Press, 1973.

Thornbury, W.D. *Principles of Geomorphology*. New York: John Wiley & Sons, Inc., 1969.

Thrower, N.J. *Maps and Man*. Englewood Cliffs, N.J.: Prentice Hall, Inc., 1972.

Turekian, K.K. *Oceans*. Englewood Cliffs, N.J.: Prentice Hall, Inc., 1976.

United Nations. *Statistical Yearbook*. Published annually.

United Nations. *World Energy Supplies*. Published annually.

United States Bureau of Mines. *Mineral Facts and Problems*. U.S. Department of the Interior, 1975.

Warren, K. *Mineral Resources*. Baltimore, Md.: Penguin Books Inc., 1973.

Wilson, C.L. (project director). *Energy: Global Prospects, 1985-2000*. New York: McGraw-Hill Book Company, 1977.

PHOTO CREDITS

Page 2, The Hale Observatories; p.3, NASA: p.31, Earthscan, Mark Edwards; p.58, National Film Board; p.79, NASA: p.87, The Winnipeg Free Press; p.125, Brazilian Embassy, Ottawa; p.126, BOAC photograph; p.127, U.S. Forest Service; p.129, Victor Zsilinszky, Timber Branch, Ontario Department of Lands and Forests; p.131, Victor Zsilinszky; p.132, Victor Zsilinzsky; p.133, U.S. Forest Service; p.134, U.S. Forest Service; p.136, U.S. Forest Service; p.137, U.S. Department of Agriculture; p.151, left, Roy W. Simonson, Soil Survey, SCS, U.S. Department of Agriculture, right, source unknown; p.153, Roy W. Simonson; p.154, Roy W. Simonson; p.166, P.D. Snavely, Jr., H.C. Wagner, N.S. MacLeod, U.S. Geological Survey; p.178, R.E. Wallace and P.D. Snavely, Jr.; U.S. Geological Survey; p.181, U.S. Geological Survey; p.183, Swissair Photo; pp.186, 187, and 188, Ray Atkeson; p.190, U.S. Coast and Geodetic Survey; p.197, Ray Atkeson; p.198, R.E. Wallace, U.S. Geological Survey; p.199, John T. McGill; p.203, U.S. Department of Agriculture, Soil Conservation Service; p.211, National Air Photo Library; p.212, U.S. Geological Survey; p.217, U.S. Geological Survey; p.221, U.S. Geological Survey; p.226, National Air Photo Library, Surveys and Mapping Branch, Dept. of Energy, Mines and Resources; p.229, National Air Photo Library; Surveys and Mapping Branch, Dept. of Energy, Mines and Resources; p.231, U.S. Geological Survey; p.233, National Air Photo Library, Surveys and Mapping Branch, Dept. of Energy, Mines and Resources; p.235, National Air Photo Library, Surveys and Mapping Branch, Dept. of Energy, Mines and Resources; p.239, Swissair photo; p.240, Swissair photo; p.243, National Air Photo Library, Surveys and Mapping Branch, Dept. of Energy, Mines and Resources; p.246, Tom Ross; p.249, National Air Photo Library, Surveys and Mapping Branch, Dept. of Energy, Mines and Resources; p.257, Spence Air Photos; p.258, Spence Air Photos; p.260, National Air Photo Library, Surveys and Mapping Branch, Dept. of Energy, Mines and Resources; p.273,

ACKNOWLEDGEMENTS

The authors wish to acknowledge the contribution of Jerry Salloum to Chapters 1, 2, 3, and the Appendix.

INDEX